T. W. Goodwin, C.B.E., F.R.S.

History of the Biochemical Society

1911–1986

The Biochemical Society • London

Published by the Biochemical Society, London 1987
© The Biochemical Society 1987

British Library Cataloguing-in-Publication Data
Goodwin, T. W.
　History of the Biochemical Society,
　1911–1986
　1. Biochemical Society — History
　I. Title　II. Biochemical Society
　574.19′2′06041　　QP511.5.G7

ISBN 0-904498-21-2

Typeset by Unicus Graphics Ltd., Horsham, West Sussex
Main text set in 11 on 12 point Times Roman
Printed in Great Britain at the University Press, Cambridge
Distributed by the Biochemical Society Book Depot, Colchester, Essex

Contents

1	The Emergence of Biochemistry in the United Kingdom	1–11
2	The Founding of the Society: Early Developments, 1911–1944	13–35
3	General Developments 1944–1986	37–76
4	The Finances of the Society 1944–1986	77–96
5	The Group Structure of the Society	97–113
6	The Society's Publications	115–145
7	International Activities of the Society	147–156
8	Professional and Educational Activities of the Society	157–172
Index		173

Acknowledgments

The Biochemical Society and the author acknowledge with thanks permission from those listed below to reproduce the photographs indicated:

Edward Leigh, Esq., Fig. 3.12
Royal Postgraduate Medical School, Fig. 3.2
The Middlesex Hospital, Fig. 7.2
The Royal Society, Figs. 1.1, 2.6
University of Edinburgh Medical School, Fig. 8.6
University of Leeds, Fig. 7.3
University of Liverpool, Figs. 2.9, 8.3

Abbreviations

ACP, Advisory Committee for Publications; BCCB, British Co-ordinating Committee for Biotechnology; *BJ*, *Biochemical Journal*; BS, Biochemical Society; *BST*, *Biochemical Society Transactions*; C.U.P., Cambridge University Press; EFB, European Federation of Biotechnology; FEBS, Federation of European Biochemical Societies; ICSU, International Council of Scientific Unions; IUB, International Union of Biochemistry; IUPAC, International Union of Pure and Applied Chemistry; MRC, Medical Research Council; PEC, PE(S)C, Professional and Educational (Sub-)Committee; RDS, Research Defence Society; UCL, University College London.

Preface

As the 75th Anniversary of the founding of the Biochemical Society approached the Society Committee decided that one way to celebrate this event would be to commission a History of the Society. Publications covering the first 38 years, by R. H. A. Plimmer, and the first 500 meetings, by R. A. Morton, were already in existence; they were published in 1949 and 1969, respectively. However, the speed with which the Society has developed, commensurate with that of Biochemistry itself, made the idea of a new anniversary History an attractive one. Consequently when I was invited to undertake this assignment I accepted with pleasure. However, I did stipulate that I should write a history covering the entire existence of the Society and not just an updating of Morton's *History*. I took this view not because of any perceived inadequacies in his or Plimmer's volume — indeed my indebtedness to them will soon become very clear to the reader; but I felt that there might be some merit in a single author history by someone who has been intimately connected with the Society for over 40 years. It remains to be seen if this view was warranted and whether the feeling of challenge and excitement which has pervaded the Society since the Second World War emerges from the mass of archival data which a formal history must inevitably contain. In addition to providing information I have tried to suggest the nature of the personalities behind the biochemists who were responsible for the Society's development by including photographs, thumbnail biographies and anecdotes; some of the latter have been 'handed down' (but I hope they are not apocryphal) and some have come directly from the 'horses' mouths'. Inevitably, my own views have occasionally obtruded and I should make it clear that they are mine and do not necessarily reflect those of the Society or, indeed, of any other member of the Society.

In order to give the financial problems faced by the Society's early officers more immediacy I have indicated in square brackets the present day equivalent of any sums of money quoted. The calculations have been based on official cost of living indices; thus the present day values are only approximations for there is no evidence that inflation in publishing and administration has been over the years the same as the average figure.

The response of colleagues and friends to my request for help and information was most valuable to me in coming to an integrated view of the development of the Society. To those for whom I have been able to make only passing reference to their contributions, I have space here merely for a generalized 'thank you': appropriate acknowledgement is given in the text and their full replies are now lodged in the Society's Archives. A recent innovation has been to appoint a Society Archivist, who is currently Professor G. R. Barker. His efforts should make life much easier for a future historian of the Society.

Help in providing written information which I have drawn on at length was given by Professor H. R. V. Arnstein, Professor H. Bradford, Professor P. N. Campbell, Professor A. C. Chibnall, Dr R. M. C. Dawson, the late Professor K. S. Dodgson, Dr D. F. Elliott, Dr D. S. Jones, Professor J. Lucy, Dr P. T. Nowell, Dr A. G. Ogston, Professor C. Pogson, Dr H. J. Rogers, Dr G. A. Snow, Professor R. H. S. Thompson, Dr D. C. Watts and Professor W. J. Whelan. A number of these have also commented on my manuscript and many sections have been improved following their informed criticism. Mr A. Allan, Assistant Archivist, The University of Liverpool, kindly answered my many queries about Chapter 1.

The Society's professional staff, with the benign encouragement of Mr Glyn Jones (Executive Secretary), have given every possible assistance. In particular I must acknowledge the help of Mr Tony Evans (Editorial Manager), who saw the book through the press and prepared the Index; his advice based on many years' experience with the Society's publications was invaluable. It was an enormous help to be able to draw on the encyclopaedic knowledge of Doris Herriott (Meetings Secretary), particularly with regard to meetings and administrative matters. I am also most grateful to Vivienne Avery, who, before she left the Society's staff, gave enthusiastic help in the early stages of the project, and Dianne Stilwell who, on appointment to a new post of Research and Information Officer, immediately plunged into archival searches and spent a great deal of time organizing the photographs.

I wish particularly to pay tribute to the help given by the late Professor K. S. Dodgson (a member of the first honours class which I taught), who was one of the initiators of the idea of a History for the 75th Anniversary of the Society and who gave it his wholehearted support before he was struck down with his last illness. Sadly he was not to enjoy any of the Anniversary celebrations, the concept of which owed so much to his enthusiastic advocacy.

I hope that I have been able to show how and why the Biochemical Society has developed into one of the most successful and effective scientific societies in existence. It shows every sign of remaining so and I am confident that my successor writing on the Centenary of the Society will be reporting a similar situation.

T.W.G.
February 1987

Chapter 1

The Emergence of Biochemistry in the United Kingdom

1.1 Introduction
1.2 Early Developments in the U.K.
1.3 The First Chair of Biochemistry in the U.K.
1.4 Early Days at Cambridge

1.1 Introduction

A new science inevitably starts with a number of innovative, imaginative and enthusiastic investigators who, trained along classical lines, break away from tradition to attack problems from a fresh viewpoint. In experimental subjects the success of the new approach frequently depends on the ingenious interpretation and development of techniques borrowed from more than one established discipline. Growth of a new science will be rather slow and sporadic until the pioneers realize that significant progress can only be achieved by collective action through some type of organized body. Such a group of like-minded scientists should be able to weld themselves together as an active force to persuade their colleagues in traditional subjects that they indeed have something worthwhile to offer; they should also be capable of dealing with the inevitable obstruction from vested interests of well-established classical disciplines, and of persuading the appropriate authorities to provide research laboratories and University departments. Eventually they should be sufficiently confident to develop their own Journal. The two disciplines which spawned Biochemistry, or if you will physiological chemistry, in the latter half of the nineteenth century and the early years of the present century were physiology and organic chemistry, and the environment in which the subject grew was generally but not exclusively within a medical school.

The major centres for biochemical research between 1840 and 1870 were the German medical schools. Before 1840 most of the chemists in German Universities were in medical faculties and in this way many of them were attracted to biological research problems. However, chemistry was feeling its feet and transferring to philosophical (science) faculties as a pure science. This, combined with the decision in 1840 of the committee concerned with reforming medical education in Germany that organic chemistry should be taught by chemists in the philosophical faculties, led to the development of organic chemistry at the expense of physiological chemistry. Many personal (*ausserordentlich*) chairs were held in medical schools by eminent physiological chemists at this stage but when they retired their chairs either disappeared or were transferred to the philosophical faculties.

Only one established (*ordinarius*) chair existed; this chair, which started as a personal Professorship in 1845 at Tübingen for J. Schlossberger, was elevated to an established chair in 1859. Schlossberger was succeeded by Hoppe-Seyler but the chair was transferred in 1863 to the philosophical faculty as a second chair in the chemical institute; however, when periodically it became vacant it continued to be filled by first-rate biochemists. Indeed it was the one chair established during this period to survive into the twentieth century. The fascinating story of the early developments of Biochemistry in Germany is described in detail by Kohler [1]. Because of these developments many German biochemists in the latter half of the nineteenth century became extremely well known for their outstanding research although they frequently did not work in a conventional biochemical environment. The pity is that they did not establish research schools which survived them; as Kohler [1] succinctly puts it, German Biochemistry represented "high intellectual achievement on the margin of the discipline and undeveloped institutions at the core".

Even the appearance of a high quality journal (*Hoppe-Seyler's Zeitschrift für physiologische Chemie*) in 1877 did not lead to a consolidation of biochemical activities in appropriate departments and the full development of the German biochemical potential had to wait until the 1950s, following the founding of the Gesellschaft für Physiologische Chemie in 1947 (the Society was re-named Gesellschaft für Biologische Chemie in 1964).

In spite of the rush of American scientists to study in Germany in the mid eighteen hundreds and the fact that American biochemical centres between 1875 and 1900 bore strong similarities to their German counterparts, the essential development of the discipline in the U.S.A. was very different from that in Germany; by 1909 some 60 out of 97 medical schools offered courses in physiological chemistry. As Kohler [1] points out, although the continuous development of Biochemistry had its problems it was well established in both medical and agricultural schools by 1939 and, as is now clear, it occupies today a pre-eminent position in American scientific circles. Again we can turn to Kohler [1] for full details of this evolution.

The early developments in the U.K., to be discussed in the following sections, rather fell between those in Germany and those in the U.S.A. Full appreciation of the significance of the subject came a little later than in Germany but by 1939 it was accepted as a respectable scientific discipline, although still only practised on a small scale. However, there was a sound foundation on which to build the relatively massive edifice of Biochemistry which has arisen in the U.K. since the end of the Second World War.

1.2 Early Developments in the U.K.

As early as 1802 Humphry Davy was lecturing on 'Agricultural Chemistry' at the Royal Institution and in 1809 The Royal Society announced the formation of a Society for the Promotion of Animal Chemistry. This was to be regarded as an Associated Society; in particular, all papers read before it were to be offered for publication in the *Philosophical Transactions of The Royal Society*. A number of papers, mainly on animal secretions, were published but sadly the Society soon became little more than a dining club and eventually disappeared [2]. However, the influence of developments in Germany was gradually having an impact and by the end of the nineteenth century physiological chemistry (Biochemistry) was appearing on the scene in medical schools, where it was taught as part of the pre-clinical instruction.

The centre of this development was University College London (UCL), where Dr W. D. Halliburton (Fig. 1.1) ran a course from 1884 to 1890 when he moved to the Chair of Physiology at King's College London. There he built up what has been described as the first research school in Biochemistry in the U.K. Certainly he seemed to be the first British scientist to visualize Biochemistry as a wide-ranging science in its own right and not merely as a handmaiden to medicine. He was a leading light in the founding of the Biochemical Society (Chapter 2). According to Gowland Hopkins "he was the first in this country, by his works and his writings, to secure for Biochemistry general recognition and respect" [3]. All these achievements were acknowledged when he was elected the first Honorary Member of the Biochemical Society. Chemical Physiology at UCL had its ups and downs after Halliburton's departure but was stabilized in 1901 by the new Professor of Physiology, E. H. Starling, who established two posts, one of which, Assistant in Physiological Chemistry, was filled in 1909 by R. H. A. Plimmer (Fig. 1.2), a protagonist in the efforts to found the Biochemical Society.

At about the same time as Halliburton's activities at King's College London (1895), Cambridge formalized Sheridan Lea's post in the Physiology Department as a lectureship in Physiological Chemistry. Lea soon had to resign owing to ill health and a crucial appointment was made when Gowland Hopkins (Fig. 1.3) filled the vacancy left by Lea's departure [1] (see section 1.4). W. Ramsden and J. S. Haldane were teaching physiological chemistry at Oxford in 1897 but the first efforts

Fig. 1.1. W. D. Halliburton, F.R.S. (1860–1931). Professor of Physiology, Kings College London 1910–1925. A pioneer of British Biochemistry who was the first Honorary Member of the Biochemical Society (1923).

Fig. 1.2. R. H. A. Plimmer. Founder member of the Biochemical Society. Honorary Secretary, 1911–1919. Chairman of the Society's Committee, 1922–1923, 1939–1940. Honorary Member, 1943.

Fig. 1.3. Sir Frederick Gowland Hopkins, O.M., P.R.S. Sir William Dunn Professor of Biochemistry, University of Cambridge 1922–1936. First Chairman of the Biochemical Society Committee, 1913–1914. Nobel Laureate, 1929. Honorary Member, 1943.

to improve the facilities for physiological chemistry did not take place until 1906, and Ramsden was not appointed to a lectureship until 1914 when C. S. Sherrington arrived on the scene from Liverpool [1] to take up the Waynefleet Chair of Physiology. Meanwhile, at Liverpool Dr A. S. Grünbaum held a lectureship in Physiological Chemistry in Sherrington's Department of Physiology from 1898 until he accepted the Chair of Pathology at Leeds in 1904. However, an imaginative and far-reaching decision made at Liverpool at that time was to establish the first Chair of Biochemistry in the U.K. Events leading to this are described in the next section.

In Scotland Physiological Chemistry began to be taught to medical students at Edinburgh soon after the arrival of E. A. Schäfer in 1899 as Professor of Physiology, when he appointed T. Milroy as lecturer in 'advanced physiology and physiological chemistry'. Biochemistry arrived a little later in Glasgow when in 1905 a bequest from Dr J. Grieve, a medical graduate of the University, was used to found a lectureship in Physiological Chemistry. The appointee, Dr E. P. Cathcart, was destined to play a most important part in the development of Biochemistry at Glasgow [4] by becoming the first incumbent of the Gardiner Chair of Physiological Chemistry within the Institute of Physiology in 1919. The Chair is named after two Glasgow ship-owning brothers who bequeathed to the University sufficient money to endow three Chairs, one of which was in Physiological Chemistry. The budding off of a separate Department of Biochemistry from the Institute of Physiology did not, however, occur until 1948 when J. N. Davidson, who played an important role in the development of the Biochemical Society, was appointed to the Gardiner Chair. Furthermore, implementation of Davidson's decision to change the name of the Chair from Physiological Chemistry to Biochemistry was delayed until 1958 because a change in a University Ordinance was required and that had further to be approved by the Privy Council. In Edinburgh, contrary to expectations, the development of Biochemistry was not made easy by the institution of a Chair of Chemistry Related to Medicine in 1929. Local University political activity resulted in the responsibility for teaching elementary chemistry to medical students being assigned to this Chair, whilst Biochemistry teaching remained in the hands of the Physiology Department. Thus no focal point existed to draw biochemical activities together and although many renowned biochemists were associated with the Chair of Chemistry Related to Medicine, Biochemistry did not free itself from Physiology until after the Second World War.

In the University of Wales, preclinical teaching in the Welsh National School of Medicine began in 1894 and the elements of Biochemistry were included in the Physiology Course, for the College's Calendar for 1894–1895 records that "the student will himself experiment with the properties of albumen and its allies, the carbohydrates and fats of the food, blood, milk, the digestive juices, glycogen and wine*". However, it was not until 1910 that a member of the Department of Physiology, R. L. Mackenzie Wallis, was given the title of lecturer in Physiological Chemistry. A separate Department of Biochemistry did not emerge from the physiological nest until as late as 1956, when John Pryde, a member of staff since 1925, was appointed to the newly created Chair of Biochemistry [5].

Agriculture has always been a major interest in two other Constituent Colleges of the University of Wales, Aberystwyth and Bangor, and Agricultural Chemistry, somewhat akin to physiological chemistry as taught in a medical school, was very early part of the curriculum for degrees in Agriculture. The first lecturer in Agricultural Chemistry at Aberystwyth was J. Jones Griffiths (later Professor of Agriculture), who was appointed in 1906. The story of the eventual emergence of Biochemistry at Aberystwyth as a distinct discipline within the milieu of Agriculture has been delicately recounted by R. J. Colyer [6]. A somewhat similar historical development at Bangor has been described with characteristic enthusiasm and bluntness by Professor W. Charles Evans [7]. An autonomous Department of Biochemistry was not established at University College, Swansea until 1972.

In Queen's University Northern Ireland the first Professor of Physiology was appointed in 1902, and a lecturer in Physiology, J. A. Milroy, appointed at the same time, was redesignated lecturer in Biochemistry in 1909. In 1924 an autonomous Department of Biochemistry was established and Milroy, by now a Reader, was appointed to the newly endowed J. C. White Chair of Biochemistry. J. C. White was a Belfast City Councillor [8].

In Trinity College Dublin a special lectureship in Biochemistry in the Department of Physiology was established in 1921. This was converted into a personal Chair for W. R. Fearon in 1934 but an independent Department of Biochemistry was not inaugurated until 1960. In the National University of Ireland Biochemistry first appeared on the scene in 1934, when a Department of Biochemistry and Pharmacology was established with E. J. Conway as professorial head [9].

From 1920 until the beginning of the Second World War the biochemical presence in British Universities increased so that by 1939 it was taught in 18 institutions, where six independent Chairs and four Dual chairs (Biochemistry with Physiology) had been established. The major expansion, however, occurred after the Second World War and in 1986 Professors of Biochemistry, in one guise or another, are found in all 44 Universities and also in many Polytechnics.

*Alas, a misprint for urine.

1.3 The First Chair of Biochemistry in the U.K.

To return to the early days, the most significant development at the turn of the century was, as indicated earlier, the decision of the University of Liverpool to establish a Chair of Biochemistry, the first in the U.K., as part of a programme to develop medical research. The leader in this move was Sir Rupert Boyce, the George Holt Professor of Pathology, who persuaded William Johnston, a wealthy Liverpool shipowner and Boyce's father-in-law, to support his proposals. Johnston first promised £5,000 towards implementing the University's decision to develop medical research but later reconsidered his offer and increased it to £25,000 [at least £750,000 at 1986 prices] in a letter written to the Principal of the University on 1 February 1902:

> "I wish this sum to be devoted to the advancement of Medical Science, and I propose, with the approval of the Council, that it shall be divided as follows:
> £10,000 to found a Chair of Biological Chemistry,
> £9,000 to be used for building purposes,
> £6,000 to be devoted to permanently endowing my two Fellowships of Colonial and International Medical Research, and for founding a third Fellowship in Gynaecological Research.
> I am desirous that the sum of £9,000 shall be made to cover all expenses of erecting and fitting the proposed building, which I suggest shall be simply constructed and designed to give a maximum amount of space to research and teaching, and to adjoin the Thompson–Yates Laboratories. I wish the building to have accommodation for research in Physiology and Pathology, for the Tropical School of Medicine, for the new Professor of Biochemistry and for Clinical Pathology."

The increased offer was stimulated by the tragically ironic death of his daughter (Boyce's wife) in childbirth. The Fellowship in Gynaecological Research mentioned in the letter still carries her name.

This munificent gift was accepted by the University Council within two days of receiving the letter and they implemented the new proposals with such speed that at a special meeting on 29 July 1902 it was possible to approve the appointment of Benjamin Moore as Professor of Bio-Chemistry (the hyphen soon disappeared) on the conditions outlined in a minute recorded in copper plate handwriting (Fig. 1.4). The appointment was specifically dissociated from the teaching of medical students, although it was located in the Faculty of Medicine. This exclusion clause was probably due to Sherrington who, although he approved of physiological chemistry, had previously laid claim to it by appointing a lecturer in the subject in the Department of Physiology in 1898, and he would not have wished to see it drawn into the new Department.

Dr B. Moore appointed to the Chair of Bio-Chemistry.

7. It was moved by the Chairman, seconded by Mr Muspratt and

Resolved

"That on the recommendation of Senate and in accordance with the provisions of the Charter and Statutes, Mr Benjamin Moore, M.A., D.Sc., Royal University of Ireland, be invited to accept the Chair of Bio-Chemistry on the terms stated in the above report, at a salary of £375 per annum and half the fees of his students. The Professor to commence his duties on the 1st of October next."

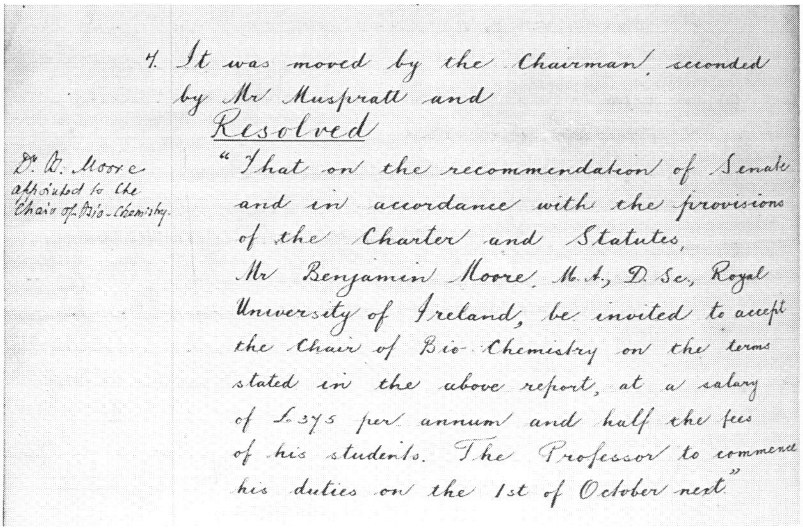

Fig. 1.4. The Minutes of the special meeting of the Council of the University of Liverpool recording the appointment of the first Professor of Biochemistry in the U.K. (Kindly provided by the Archivist of the University of Liverpool.)

The Chair was essentially a research Chair, with which was combined some advanced teaching, and although it was clearly intended that the holder should orientate his research in a medical direction Moore was given the freedom to develop Biochemistry relatively unrestricted as a true discipline within the Life Sciences. More importantly, he was freed from the shackles of physiology.

The salary arrangements offered to Moore, £375 p.a. [*c.* £12,000 today] plus half the fees of his students, would be readily acceptable to heads of large departments if 'students' were defined as the University Grants Committee's FTEs (full time equivalents).

Moore (Fig. 1.5) was a quite extraordinary person; he graduated in Engineering at Queen's College (now University) Belfast but then moved to Leipzig to work in Ostwald's laboratory, from where he moved to UCL to join the Physiology Department under Sharpey-Schafer. This was followed by a period at Yale after which he was appointed lecturer in Physiology at Charing Cross Hospital Medical School. During this period he qualified in medicine [10]. In 1902, as just recorded, he was appointed to the Johnston Chair of Biochemistry. With his great breadth of experience, his wide vision and his remarkable energy he collected around him able and enthusiastic assistants and colleagues and together they published on, *inter alia*, diabetes, photosynthesis, renal calculi and heavy metal toxicity. Amphibia and marine algae also came under Moore's scrutiny but perhaps his most important work was on membranes: thus Gowland Hopkins, in an obituary of Moore [10], said "it is clear for instance, that he possessed at this time (1910) a fairly definite conception of the

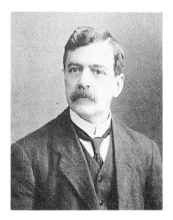

Fig. 1.5. Benjamin Moore, F.R.S. Johnston Professor of Biochemistry, University of Liverpool 1902–1914. Founder of the *Biochemical Journal.* Whitley Professor of Biochemistry, Oxford 1920–1922.

Fig. 1.6. Edward Whitley, M.A., Benjamin Moore's wealthy research assistant who helped him launch the *Biochemical Journal* in 1906. He also provided the funds to found the Whitley Chair of Biochemistry at Oxford. (Photograph kindly provided by Dr T. Moore, Benjamin Moore's son, who obtained it from Mr E. Whitley, grandson.)

membrane equilibrium which four years later, was quantitatively studied and clearly defined by Donnan".

One of Moore's assistants was Edward Whitley (Fig. 1.6), a wealthy Oxford graduate in physiology and psychology. He supported Moore financially when he decided to found the *Biochemical Journal* in 1906. He was the son of Edward Whitley M.P., sometime Lord Mayor of Liverpool, and his wealth came from a family connection with the brewers Greenall Whitley. Whitley did Biochemistry another signal service when in 1920 he provided the University of Oxford with £10,000 [£200,000] to endow the Whitley Chair of Biochemistry [11]. It seemed entirely fitting that the first holder of this Chair should have been his old friend and colleague Benjamin Moore. Sadly Moore died in 1922 before he had time to stamp his personality on the Oxford scene. One can only regret the loss suffered by both British Biochemistry and Oxford by the early death of this imaginative and impressive man.

Moore's breadth of vision resulted in the introduction at Liverpool in 1910 of the first Honours School of Biochemistry in the U.K., and his achievements must have impressed on Sherrington that the future of Biochemistry lay outside a Physiology Department, because just before he (Sherrington) left Liverpool for Oxford in 1913 he persuaded Moore to take over the teaching of Biochemistry to medical students.

Unfortunately Moore, who had been elected a Fellow of The Royal Society in 1912, left Liverpool in 1914, soon after Sherrington, to enter the newly established National Institute for Medical Research at Hampstead. His successor, Walter Ramsden (Fig. 1.7), did not seem to possess Moore's all-embracing view of Biochemistry and had apparently neither the drive nor interest to develop Moore's imaginative ideas. In fact, his early promise in research did not materialize. Ramsden, who was, however, active in founding the Biochemical Society, was mainly interested in Medical Biochemistry and he cultivated the clinical aspects of the subject. Thus Biochemistry at Liverpool gradually reverted to its traditional role of handmaiden to Physiology and Medicine. Indeed a proposal from the Medical Faculty that the name of the Department be changed to Chemical Physiology was fortunately defeated. The situation did not change until the appointment of H. J. Channon in 1932, when Ramsden, a bachelor, retired to return to Oxford to continue his Sheppard Medical Fellowship of Pembroke College, which was 'tenable for life unless forfeit by marriage'. Channon was given the specific remit to develop Biochemistry as such and to revitalize the Honours School. This he did with some vigour.

Whilst the broad view of Biochemistry as a primary discipline was disappearing temporarily from the Liverpool scene it was being pursued with great fortitude by Gowland

Fig. 1.7. Walter Ramsden. Founder member of the Biochemical Society. Johnston Professor of Biochemistry, University of Liverpool 1914–1931.

Hopkins (Fig. 1.3) at Cambridge. However, he found the going heavy at times.

1.4 Early Days at Cambridge

The Professor of Physiology (M. Foster) at Cambridge in the 1880s began to develop Chemical Physiology and, as indicated earlier, Sheridan Lea who was responsible for this teaching was appointed to a special lectureship in 1895. When Lea retired owing to ill health, the position was not immediately continued but by 1898 it was revived with no stipend attached. £100 was allocated temporarily from departmental funds and on this basis, with a supplement for teaching at Emmanuel College, Hopkins at the age of 38 succumbed to Foster's entreaties and accepted the post. This type of administrative manoeuvring was to dog Hopkins until the 1920s when the trustees of the Dunn bequest made a massive contribution to his department. Hopkins had the same view of Biochemistry, a fundamental subject central to biology, as had Halliburton and Moore, and with great determination, wide vision and a magnificent command of the English language (there can be few biochemists who have in their undergraduate years managed to avoid writing an essay on one or other of Hopkins's famous aphorisms), he eventually achieved the acceptance of Biochemistry as a Part II (Honours) subject during the First World War and collected around him a group of outstanding research workers.

Intertwined with the development of the science of Biochemistry at Cambridge was Hopkins's further aim of establishing an institution free from the (generally) benevolent patronage of the Physiology Department. The love–hate relationship between the physiologists and the protagonists of the rapidly developing science of Biochemistry was, to some extent, the basis of the difficulties which Hopkins faced, but more important were the labyrinthine activities of the University and the Colleges; furthermore Hopkins's personality apparently did not allow general administrative activities to become easy to him; he was "shy, diffident to a fault, and at a loss in the rough and tumble of University politics" [1]. His problems have been frequently described [12] but it is worth outlining them here in order to see how much he had to endure.

In 1902 he apparently refused the Liverpool Chair [13] on the understanding that he would be offered a readership at Cambridge; however, no such offer exists in the Liverpool archives. In the event, Hopkins's Professor (J. Langley) did not recommend an endowed readership but an upgrading of his title to Reader with an increased contribution of the University to his salary (£50–£100) [£1800–£3600]. However, Langley supported Hopkins more strongly when he felt that he was not in direct competition with the Physiology Department for funds. Kohler [1] gives two examples. On the one hand, there was the strong support Langley gave to the attempt in 1906 to obtain the Quick Chair of Biology for Hopkins. This foundered because according to W. M. Fletcher (eventually, the first secretary of the Medical Research Council) "the interference of lay opinion outside, unskilfully handled, led to its association with a science whose importance is secondary and not primary" [1]. On the other hand, when a Draper's bequest of £2200 [£26,000] for a new laboratory for Physiology came along in 1909, Hopkins's hope for 'self-contained and independent laboratories' was dashed when the money was directed elsewhere.

Meanwhile, in 1906 Emmanuel elected Hopkins Science Tutor, which also involved a Fellowship which he held until 1910. By then Fletcher had entered the fray both for Hopkins and the maintenance of the status of Cambridge Physiology/Biochemistry. He was stimulated into activity by the decision of the University Senate that it could not support a Chair in Biochemistry without outside help: "although Foster brought the centre of gravity of English Physiology to Cambridge, and left it there, it has been seriously displaced recently by the foundation of a chair of Biochemistry at Liverpool and the publication there of the *Biochemical Journal*" [13]. This is an extract of part of a document presented to Trinity College Council recommending the election of Hopkins to a Fellowship and Praelectorship, tenable with the University Readership. The recommendation was accepted by the College in the "confident hope that the University will not relax its efforts to secure at the earliest opportunities the foundation of a Professorship created without salary or endowment" [14]. This was eventually implemented in 1914 but there was still no possibility to develop the study of Biochemistry along the lines Hopkins wished for. Eventually his dreams were realized in 1919 by a magnificent bequest of £210,000 [£4.5 × 10^6] by the Dunn trustees to set up a Dunn Biochemistry Institute. This Institute was opened in 1924 and Hopkins, now holding an endowed Chair (the Sir William Dunn Chair of Biochemistry), was poised to achieve his ambition — the rest is history.

The procrastination at Cambridge meant that Hopkins's Chair was only the third to be established in the U.K. The second was established in 1912 by the University of London at the Lister Institute in order to recognize the outstanding research achievements of Sir Arthur Harden (Fig. 1.8). Harden was also outstanding in his service to the Biochemical Society; this will be described in the next chapter.

Although the mainstream of Biochemistry evolved from medical school-teaching in Universities, it must be remembered that qualitatively impressive contributions although on a small scale were made by the fermentation industries and agri-

Fig. 1.8. Sir Arthur Harden, F.R.S. Professor of Biochemistry, Lister Institute 1912–1930. Founder member of the Biochemical Society. Chairman of the Society's Committee, 1914–1915. Editor of the *Biochemical Journal*, 1913–1937. Nobel Laureate, 1929. Honorary Member, 1938.

cultural science. This as we shall see was appreciated by the founders of the Biochemical Society. Biochemists were also appearing in departments of botany, zoology and pathology and in retrospect it is clear that even as early as 1900 the writing was on the wall that a new central discipline within biology was emerging.

With the background summarized in this chapter, we can now move to consider the formation of the Biochemical Society in 1911 and appreciate how the founders got the timing exactly right.

References

1. Kohler, R. E. (1982) *From Medical Chemistry to Biochemistry*, Cambridge Monographs on the History of Science. pp. 399. Cambridge University Press, Cambridge.
2. Hall, M. B. (1984) *All Scientists Now.* pp. 261. Cambridge University Press, Cambridge.
3. Morgan, N. (1983) *William Dobinson Halliburton F.R.S. (1860–1931). Pioneer of British Biochemistry?* Notes and Records of The Royal Society, **38**, 129–145.
4. Thomson, R. Y. & Smellie, R. M. S. (1983) Biochemistry. *Proc. R. Soc. Edinburgh, Sect. B*, **84**, 21–35.
5. Dodgson, K. S. (1985) Manuscript deposited in the Biochemical Society's Archives.
6. Colyer, R. J. (1982) *Man's Proper Study.* pp. 182. Gomer Press.
7. Evans, W. C. (1985) Manuscript deposited in the Biochemical Society's Archives.
8. Elmore, D. T. (1985) Manuscript deposited in the Biochemical Society's Archives.
9. Harrington, M. G. (1985) Manuscript deposited in the Biochemical Society's Archives.
10. Hopkins, F. G. (1927) Benjamin Moore 1867–1922. *Proc. R. Soc. London, B*, **101**, xvii–xix.
11. Morton, R. A. (1972) Biochemistry at Liverpool — 1902–1971. *Med. Hist.*, **16**, no. 4.
12. Needham, J. & Baldwin, E. (1949) *Hopkins and Biochemistry.* pp. 361. W. Heffer, Cambridge.
13. Fletcher, W. (1910) Professorial Fellowship and the Needs of the University, Trinity College (Cambridge) Council Reports, 20 May 1910*.
14. Trinity College (Cambridge) Council Minutes, 27 May 1910*.

*Reports kindly provided by Sir Andrew Huxley, O.M., F.R.S., Master of Trinity College.

Chapter 2

The Founding of the Society: Early Developments 1911–1944

2.1 Formation of the Biochemical Club
2.2 Acquisition of the *Biochemical Journal*
2.3 Emergence of the Biochemical Society
2.4 Financial Position of the Society
2.5 General Developments
2.6 Honorary Members
2.7 Discussion Meetings
2.8 Proceedings

2.1 Formation of the Biochemical Club

The events outlined in Chapter 1 which occurred in the first decade of this century made it clear that British biochemists needed a separate forum where they could develop their subject on a national level. The general criteria necessary for establishing a new discipline, summarized at the beginning of Chapter 1, were clearly already achieved.

The time was thus ripe for the formation of a Society devoted to the furtherance of Biochemistry and, on 16 January 1911, J. A. Gardner (Fig. 2.1) and R. H. A. Plimmer (Fig. 1.2), after preliminary discussions with close colleagues, sent out invitations to fifty persons likely to attend a meeting to be held at the Institute of Physiology, UCL at 2.30 p.m. on Saturday, 21 January 1911, to consider the formation of a Biochemical Society. Plimmer was evidently stung into action by an article in the press describing a new science, Biochemistry, which was making rapid progress on the Continent but was apparently unknown in Britain. The invitation, written on a postcard, read

> "Numerous suggestions having been made that a Biochemical Society should be formed in the Country, we shall be glad if you could make it convenient to attend a meeting at the Institute of Physiology, University College London, on Saturday 21st January at 2.30 p.m. to discuss the question."

Today five days would seem very short notice for a meeting — the postcard might even have not arrived in time — but perhaps the diaries of senior biochemists did not fill up quite so quickly in 1911 as they do in 1986. It is also interesting that Saturday afternoon was then considered an acceptable time for a meeting. In fact thirty-two attended the meeting and a further fourteen sent encouraging replies.

Fig. 2.1. J. A. Gardner. Founder member of the Biochemical Society. Honorary Treasurer, 1913–1943.

In the words of R. H. A. Plimmer, who wrote the first history of the Society from 1911 to 1949 [1]:

"J. A. Gardner presided and gave the chief reasons for calling the meeting.* He emphasized the growing importance of Biochemistry both on the animal and vegetable sides. The increasing number of workers rendered the formation of a Biochemical Society desirable for four main reasons: (1) a common meeting place to discuss biochemical problems; (2) the association of the workers on the animal and vegetable sides; (3) a common journal to be owned by the Society; (4) the advancement of Biochemistry in this country.

"Professor W. D. Halliburton, in opening the discussion, was strongly in favour of the formation of such a combined society with its meetings on unconventional lines. He moved a resolution to this effect, which was seconded by Doctors F. G. Hopkins, A. E. Garrod and A. Harden.

"Professor H. E. Armstrong, who was opposed to any specialization, said that the main object should be to have a 'focus point', and that a society or club wherein the social side of the gathering preponderated should be a primary condition. Dr E. J. Russell, speaking for agriculture, said the number of scientific papers was not large, and thought they would be of more value if brought before other biochemists. Dr E. F. Armstrong hoped no omission would be made of workers on the botanical side. Dr Plimmer, in summarizing the subjects so far mentioned said that the chemistry of brewing came into consideration as well.

"Finally, it was proposed by H. E. Armstrong, seconded by W. D. Halliburton and carried unanimously, 'that provisionally a club be established to promote intercourse among those biologists and chemists who are mutually interested and concerned in the investigation of problems common to biologists and chemists.

"To make preliminary arrangements Professor Halliburton proposed that there should be a small committee limited to the conveners of the meeting. As these two gentlemen did not sufficiently represent all the interests, a committee of five was chosen: Gardner, Garrod, Halliburton, Plimmer and Russell. Plimmer was asked to be Secretary."

After immediately circulating interested parties that a Biochemical Club or Society was to be formed, the Committee drew up provisional rules and regulations, which were relatively informal and based on those of the Physiological Society, and called a meeting at UCL for 4 March 1911. Seven communications were presented at the meeting, after which thirty-eight members adjourned for dinner and a consideration of the proposed rules. After prolonged and vigorous discussion the rules were accepted with two amendments: (i) that the group be provisionally named 'The Biochemical Club' and (ii), following consideration of a letter from a lady who wished to

---

*In the minutes recorded as "Mr Gardner sitting on the table made some observations".

become an original member, that only men be eligible for membership. Both these amendments were destined to be revoked. The second bizarre decision (carried by 17 votes to 9) was soon challenged and at a Committee meeting on 13 July 1912 it was reversed by 24 votes to 7 and at the first meeting at which new members were elected (5 February 1913) three of the seven new members were women; they were Dr Ida Smedley — later Dr Smedley-McLean, the first woman Chairman of the Committee (1927), Dr (later Dame) Harriet Chick and Muriel Wheldale. The designation 'Club' was considered more appropriate to a group without its own scientific Journal. The saga of the acquisition of the *Biochemical Journal* is discussed later (section 2.2).

The honour of being the first fully paid-up member of the Biochemical Club was claimed by (Sir) Charles Lovatt Evans (UCL) (see Fig. 2.13), who recalls that he paid Plimmer in his laboratory immediately after the first meeting.

At the first meeting a relatively large Committee (14) was set up to function for 1911–1912. This size was necessary so that all aspects of Biochemistry would be adequately represented. The constitution of the original Committee with the members' affiliations is given in Table 2.1. The indefatigable Plimmer acted both as Honorary Secretary and Honorary Treasurer. It is significant that right at the outset the Society cast its net as widely as possible. This policy has continued throughout the years partly deliberately and partly owing to the irresistible pressure of a buoyant and expanding science. There is no doubt that this has been the correct approach and that it accounts in great measure for the scientific strength of the Society today. The location of the meetings in 1911–1912 (Table 2.2) emphasizes the wide range of interests wisely cultivated by our founders.

The visit to Rothamsted on 10 June was particularly memorable in that the members were shown, amongst other

Table 2.1. The Founding Committee of the Biochemical Club, 1911–1912

Name	Institution
H. E. Armstrong, F.R.S.	City & Guilds College (eventually Imperial College)
W. M. Bayliss, F.R.S.	U.C.L.
A. J. Brown, F.R.S.	University of Birmingham
H. H. Dale, F.R.S.	University of London
J. A. Gardner	University of London
A. E. Garrod, F.R.S.	U.C.L.
W. D. Halliburton, F.R.S.	King's College London
A. Harden, F.R.S.	Lister Institute
F. G. Hopkins, F.R.S.	University of Cambridge
F. Keeble, F.R.S.	University of Reading
B. Moore, F.R.S.	University of Liverpool
W. Ramsden	University of Oxford
E. J. Russell, F.R.S.	Rothamsted Experimental Station
R. H. A. Plimmer	U.C.L.

Table 2.2 Venues of the Meetings of the Biochemical Club, 1911–1912

Date	Location
4 March 1911	U.C.L.: Physiology Department
6 May 1911	Oxford
10 June 1911	Rothamsted Experimental Station
4 July 1911	City & Guilds College, South Kensington*
14 October 1911	School of Agriculture, Cambridge
17 November 1911	King's College London: Physiology Department
12 December 1911	Lister Institute, London
3 February 1912	St Bartholomew's Hospital, London: Department of Chemical Pathology
2 March 1912	U.C.L.: Physiology Department

*Now Imperial College.

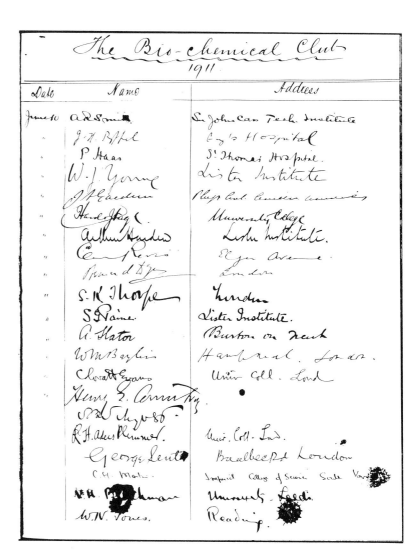

Fig. 2.2. Signatures of some of those attending the third meeting of the Biochemical Club at Rothamsted Experimental Station, 10 June 1911. (Reproduced from the Visitors' Book at Rothamsted by kind permission of the Director, Sir Leslie Fowden, F.R.S.)

things, the platinum dishes, about the size of shovels, in which Lawes and Gilbert ashed their pigs in their classical experiments. The first page of the entry in the Rothamsted visitors' book relating to this the third meeting of the Society is reproduced in Fig. 2.2.

The first Annual General Meeting of The Biochemical Club was held on 2 March 1912 at UCL. The policy of holding the A.G.M. at UCL continued for over fifty years but since 1968 the location of the meeting has become decentralized. The first A.G.M. outside London was, appropriately, held in Liverpool in 1968.

At the first A.G.M., which was chaired by W. M. Bayliss (Fig. 2.3), it was reported that at the eight meetings held during the year, forty-five communications were presented to an average audience of about forty; the best attendances were at Cambridge and the Lister Institute. The attendances at the dinners were unsatisfactory. The balance sheet (Table 2.3)

Fig. 2.3. Sir William Bayliss, F.R.S. Founder member of the Biochemical Society. Chairman of the Society Committee, 1914–1915, 1919–1920. Editor of the *Biochemical Journal*, 1913–1924.

Table 2.3 Balance sheet of the Biochemical Club, 1911

Income		Expenditure	
	£ s. d.		£ s. d.
Subscriptions —		Bank Charges	0 1 6
132 Members at 10/6 + 3d	69 6 3	" "	0 2 3
Interest on deposit	0 7 4	Stationery	0 12 4
		Meetings (Tea & Servants)	8 1 0
		Printing	9 15 3
		Postage	4 12 6
		Deposit	40 0 0
		Balance	6 8 9
	£69 13 7		£69 13 7

showed an income of nearly £70 and a balance £6 8s 9d [about £200 today]. The members of the Society numbered 132. All these figures should be contrasted with those for 1984–1985 to emphasize the enormous development of the Society. During the business meeting a squabble arose about the composition of the Committee, which resulted in H. E. Armstrong (Fig. 2.4) resigning his membership and banging out of the room. "This was unfortunate as he did so much to start the club" [2]. Morton [3] recalls that "Armstrong was able and influential and could be cantankerous. The present writer heard him, as an old man, fulminating about what he felt was the dreary lack of style in the *Journal of the Chemical Society*. Armstrong ferociously enjoyed being (partly) right on many issues".

2.2 Acquisition of the *Biochemical Journal*

At meetings of the Physiological Society around the turn of the century, the number of biochemically orientated papers

Fig. 2.4. Professor H. E. Armstrong, F.R.S. Founder member of the Biochemical Society. Prominent in the debate over 'Club or Society?'.

presented often crowded out those concerned with pure physiology. If this had continued it would certainly have strained the traditionally good relationship between biochemists and physiologists. Chemical papers produced by biochemists were published in the *Journal of the Chemical Society* but as chemists generally considered biochemists to be physiologists there was no satisfactory outlet for physiologically orientated chemical papers except for the privately run *Biochemical Journal* (see later), which, in any case, was founded only in 1906. The situation in the U.K. contrasted with that in Germany where *Hoppe-Seyler's Zeitschrift für physiologische Chemie* began in 1877 and *Hofmeister's Beitrage* in 1901; these were followed in 1906 by the *Biochemische Zeitschrift*. In the U.S.A. the *Journal of Biological Chemistry* was founded in 1905.

It was an important and immediate aim of the Biochemical Club to develop its own Journal as a mandatory step to becoming a *bona fide* scientific society. The achievement of this aim was not as straightforward as might have been thought because the *Biochemical Journal* already existed. It had been founded in Liverpool by the irrepressible Benjamin Moore with the financial help of his co-worker and patron E. Whitley (see Chapter 1), who also helped with the editing. At the start, the *Biochemical Journal* was mainly a house journal founded because Moore was finding acceptable outlets for the research papers of him and his colleagues increasingly difficult to secure. It is fascinating to read the contents list of the first issue of volume 1 (Table 2.4). However, the *Journal* quickly widened its clientele and expanded its circulation, so that it had 170 subscribers when it was taken over by the Biochemical Club in 1912.

The financial arrangements with the publishers, The Liverpool University Press, are not clear but the *Journal* must have eventually made some profit. Dr T. Moore (Benjamin Moore's son) recalls that "from the death of my father in 1922 until well after I came to Cambridge (in 1925) I used to be greatly helped in my penury by small royalties from the L.U.P. relating to the first four (six?) volumes of the *B.J.*" [4].

The story of this takeover of the *Biochemical Journal* with its delicate and complicated negotiations between strong characters makes fascinating reading as described by Plimmer [1], who was a protagonist in this affair.

"Professor Moore was a member and strong supporter of the Biochemical Club. The Committee met Professor Moore in consultation on 11 February, 1911. It decided not to issue printed proceedings for distribution at the meetings. Professor Moore offered to accept papers of members of the Club and act in conjunction with the Committee in regard to their publication and proposed to issue the new volume under the editorship of B. Moore and E. Whitley with the collaboration of the Committee

Table 2.4. Part of Contents of Volume 1 of the *Biochemical Journal*

	PAGE
The Oxygen Tension in the Submaxillary Glands and certain other tissues. By Joseph L. Barcroft, M.A.	1
A method for determining the total daily gain or loss of fixed Alkali, and for estimating the daily output of Organic Acids in the Urine, with applications in the case of 'Diabetes Mellitus.' By Edward S. Edie, M.A., B.Sc. (Edin.), Carnegie Research Scholar, and Edward Whitley, M.A. (Oxon.)	11
On the treatment of 'Diabetes Mellitus' by acid extract of Duodenal Mucous Membrane. By Benjamin Moore, M.A., D.Sc., Johnston Professor of Bio-chemistry, University of Liverpool; Edward S. Edie, M.A., B.Sc. (Edin.), Carnegie Research Scholar, and John Hill Abram, M.D. (Lond.), M.R.C.P., Honorary Physician, Royal Infirmary, Liverpool	28
The Physiological Properties of 'West African Boxwood.' By R. J. Harvey Gibson, M.A., F.L.S., Professor of Botany in the University of Liverpool	39
Filtration as a possible Mechanism in the Living Organism. By Leonard Hill, M.B., F.R.S., Lecturer on Physiology, London Hospital, Medical School	55
The Pharmacological Action of Digitalis, Strophanthus, and Squill on the Heart. By G. S. Haynes, M.B., B.C.	62
The Action of Acids and Alkalies, and of Acid, Alkaline, and Neutral Salts upon the Tadpole of *Rana Temporaria*. By Herbert E. Roaf, M.D. (Toronto). British Medical Association Research Scholar, and Edward Whitley, M.A. (Oxon)	88
Observations on Fehling's Test for Dextrose in Urine. By Hugh MacLean, M.D., Senior Assistant in Physiology, University of Aberdeen	111
Studies in the Chemical Dynamics of Animal Nutrition. By S. B. Schryver, D.Sc., Ph.D., Lecturer on Physiological Chemistry to University College, London.	123
On the Influence of Calcium Salts upon the Heat-Coagulation of Fibrinogen and other Proteids. By Charles Murray, M.A., M.D., D.P.H., formerly Senior Assistant in Physiology in the University of Aberdeen	167
On some Aspects of Adsorption Phenomena, with especial reference to the Action of Electrolytes and to the Ash-Constituents of Proteins. By W. M. Bayliss, D.Sc., F.R.S., Assistant Professor of Physiology in University College, London	175
A Colour Reaction of Formaldehyde with Proteids and its Relation to the Adamkiewicz Reaction. By Otto Rosenheim, Ph.D.	233
Glycocoll and Total Mono-Amino-Acids in Pathological Urines. By I. Walker Hall, M.D., Professor of Pathology, University College, Bristol; Pathologist to the Bristol Royal Infirmary.	241
The Influence of X-Rays on the Nitrogenous Metabolism and on the Blood in Myelogenous Leukaemia. By Owen T. Williams, M.D., B.Sc. (Lond.), Medical Registrar to the Liverpool Royal Infirmary	249
Secretion by the Renal Tubules in the Frog. By F. A. Bainbridge, M.A., M.D., Gordon Lecturer in Experimental Pathology, Guy's Hospital; and A. P. Beddard, M.A., M.D., Assistant Physician to Guy's Hospital.	255

of the Biochemical Club. The subscription to the Club should not include the Journal, but members would be able to obtain it through the Club at a discount of 15% on the published price. These terms were reported at the meeting of the club on 4 March, 1911.

"This proposal was not acceptable to the Club committee which wanted a Journal of its own. Professor Moore was to be asked on what terms he would hand over the *Biochemical Journal* to the Club. He met J. A. Gardner and R. H. A. Plimmer on 4 July, 1911 and explained that he started his Journal because of his desire that contributions should be published as submitted without criticism or editorial suggestions. His view was that authors of poor papers would take the blame and not the Journal. He was prepared to transfer his Journal on this basis of free and unrevised publication. The cost of publication was about £150 a volume, and there was a deficit of about £200 which might be settled satisfactorily. Gardner and Plimmer pointed out that a rival journal would compete with Moore's journal and had a good chance of success as most workers in Biochemistry had joined the Biochemical Club; yet it might not succeed. Moore wrote four days later (8 July) to say that the Club should start its own journal, and in order to give the Committee freedom of action he resigned his membership of the Club.

"The Committee on 8 July, 1911 discussed the pros and cons of publishing. Some journals had guarantors who had never been called upon. It was believed it would be possible to publish a journal without loss. So Dr Ramsden (Fig. 1.7) was asked to make inquiries at the Oxford University Press, Dr Hopkins at the Cambridge University Press, Dr Plimmer at the London University Press and at some private publishers and printers. They reported to the Committee on 14 October, 1911. Comparison of the estimates showed the cost to be from £170 to £200 a volume. A private publisher offered to take the whole responsibility without guarantee and give half the profits to the Biochemical Club.

"A suggestion of H. E. Armstrong that he with Plimmer and others act as guarantors, so that a journal be speedily published, and hand over the journal when published to the Club was not received favourably.

"Professor F. Keeble (Fig. 2.5) then moved that the Journal be published by a University Press, and that detailed particulars be obtained from the Oxford and Cambridge University Presses. Doctors Ramsden and Hopkins were asked to continue their previous negotiations.

"Dr Ramsden and the Oxford University Press felt that in the interests of Biochemistry in this country two journals should not exist, and Ramsden again tried to get Moore's co-operation. Professor Moore sent a draft memorandum of his terms: a sum of £260 [£8600] payable in four yearly instalments of £65 [£2150], Professor Moore and E. Whitley to remain as editors until the money was paid. The *Biochemical Journal* had 170 subscribers of whom twenty-four were members of the Biochemical Club. The Club Committee was told later that the price represented $1\frac{1}{2}$ years' purchase at £1 1s a subscriber. This high price could not be accepted by the Committee.

"Professor H. E. Armstrong, though he considered it desirable to buy the *Biochemical Journal*, said that no more than £100 should be offered. Later at Dr Hopkins's suggestion, he proposed that Principal Miers of Manchester University be asked to assess the value of Moore's Journal to the Club. Moore and Whitley met

Fig. 2.5. Professor F. Keeble, F.R.S. Founder member of the Biochemical Society. Prominent in the acquisition of the *Biochemical Journal*.

the Committee and agreed to the valuation, but neither side was to bind itself to accept. Principal Miers agreed to act if a short statement of the negotiations with Moore were submitted to him. His valuation of £150 [£5000] was reported to the Committee on 20 January 1912. Dr Ramsden was not content with this verdict and asked Moore to send his own statement to Principal Miers. He made no alteration in his valuation. The disparity was great and not pleasing to Moore.

"The Secretary reported to the Committee that he had met Professor Moore in December and asked him if he would agree to a valuation by Mr W. M. Meredith of Messrs Constable and Co. The answer was 'Yes'. Mr Meredith had agreed to act only if he could ask any questions, and that his award was adhered to by both parties. Moore wrote that he could not agree to the conditions.

"Finally, at this meeting of the Committee, to overcome this deadlock Professor Keeble proposed that Principal Mier's valuation of £150 [£5000] for the purchase of all rights in the *Biochemical Journal* as specified in the memorandum below be communicated to Professor Moore as a definite minimum proposal from the Biochemical Club — 'Should he be unwilling to accept the proposal, the offer is made to refer to Mr Meredith for final adjudication, both parties agreeing to accept Mr Meredith's valuation as final'.

"*Memorandum*

1. In consideration of the terms contained in subsequent paragraphs the vendors, Messrs B. Moore and E. Whitley and the University Press of Liverpool, agreed to hand over the *Biochemical Journal* to the Purchasers, the Biochemical Club, as a going concern and free from all debts together with a list of subscribers thereto standing at present at 170, but all copies of back volumes and numbers already issued of the current volume shall remain the property of the vendors.
2. The Biochemical Club agreed to pay forthwith to the vendors the sum of £150 [£5000] in purchase of the goodwill and subscription list mentioned in clause 1 and to take over and be financially responsible for the issue and management of the Journal as from a date to be agreed upon.
3. The *Biochemical Journal* shall be wholly and solely held, edited and managed by the Biochemical Club.

"If this offer now made to Professor Moore be not accepted the Biochemical Club proceeds to establish its own Journal independently.

"On February 3, 1912 two letters from Professor Moore stated that he agreed to accept the Biochemical Club's proposal to buy the Journal for a minimum price of £150, but he wished for an interview with Mr Meredith to see if the figure could be raised. He desired (1) to have the first option of recontinuing the *Biochemical Journal* if for any reasons the Biochemical Club ceased to publish it, and (2) that the title should not be changed and the volumes renumbered as from the taking over. Professor Moore was informed that the Committee could not accept the limitations of the second point, and that he should give a statement that he agreed to the original terms.

"On March 2, 1912 the Secretary informed the Committee that Professor Moore was unable to meet Mr Meredith and had written agreeing to accept the valuation of £150.

"The arrangements were thus at last complete, and it was decided to take over at the completion of the current volume (number 6). Professor Moore would state in his next number that in future the *Biochemical Journal* would be issued by the Biochemical Club.

"Later, he inserted a slip repeating this information, setting out the objects of the Biochemical Club, and stating that the subscription was 25s per annum including the Journal for the year. Other subscribers were asked to pay £1 12s per volume.

The Payment of £150

"Before the negotiations with Professor Moore were completed a generous gift of £25 from Professor Sir William Osler (through Dr Ramsden) was gratefully accepted. There was a deposit of £40 and a balance of £6 from the first year. A similar balance of £40 was expected from the second year. The Secretary felt that members would like to give donations and feel that they had helped to buy the Journal for the Club. In this way £30 was subscribed. A gift of £5 from Mrs Herter was sent from New York through Dr H. D. Dakin. Dr Vincent kindly contributed the last £10.

"An agreement for the purchase was drawn up by a solicitor. The Chairman of Committee (Dr A. Harden) and the Secretary (Dr R. H. A. Plimmer) was authorized to sign the deed of assignment, and the Secretary was authorized to pay £150 [£5000] to Professor Moore and Mr E. Whitley.

"The Biochemical Society and the *Biochemical Journal* are now so well and firmly established and taken for granted that few of the present members know anything of the troublesome negotiations which harassed the Committee of the Biochemical Club during the first two years.

Final Arrangements

"Some additional details were still necessary. A subcommittee consisting of J. A. Gardner, A. Harden, F. G. Hopkins and the Secretary was appointed to report on (1) the title and constitution of the association, (2) the cost of publication of the *Biochemical Journal*, (3) the amount of subscription, based if necessary on a canvas of members.

"The subcommittee, in view of past argumentative discussions at annual meetings, decided to take a poll by postcard on three questions:

(1) Is it your opinion that membership of the Club should involve compulsory subscription to the *Biochemical Journal*?
Answer: Yes 65; No 25.
(2) In the event of the subscription to the *Biochemical Journal* not being compulsory for all members, are you prepared to

subscribe to the Journal at a cost of 15s to £1 per annum, in addition to the present subscription to the Club?
Answer: Yes 72; No 19.
(3) Are you in favour of changing the name of the association to 'The Biochemical Society'?
Answer: Yes 79; No 10.

"It was clear that the subscription to the *Biochemical Journal* should be compulsory for all members and that the title should be The Biochemical Society.

"The tenders for printing the *Biochemical Journal* showed that the most favourable terms were those of the Cambridge University Press: £200 approximately for an issue of 500 copies of eight parts of 80 pages per volume in the style of the present Journal. It was estimated that a subscription of £1 per member would cover the cost of publication. Under the title of the Journal the words 'edited for the Biochemical Society' should be inserted.

Fig. 2.6. H. W. Dudley, F.R.S. Honorary Secretary, 1922–1924. Chairman of the Society Committee, 1925–1926. Editor of the *Biochemical Journal*, 1924–1930.

Editorship of the Biochemical Journal

"The Committee decided that the *Biochemical Journal* should be edited by two editors and a representative editorial Committee.

"No definite record exists of how the first editors were chosen. The Secretary well remembers how he thought that one editor might represent the more chemical side and the other the more physiological, and that if he could secure the services of Dr A. Harden and Dr W. M. Bayliss as editors the greatest benefits would come to the Biochemical Society and the *Biochemical Journal*. He made special visits to Doctors Harden and Bayliss and was agreeably surprised and overwhelmed with delight to learn that both would accept. It was the finest possible culmination to all the work in connection with the *Biochemical Journal*."

The choice of Harden and Bayliss as editors was an inspired one; they worked together until 1924, during which time the high standards always associated with the *Journal* were firmly established. Harden, however, carried on when Bayliss retired and with a succession of assistants, H. W. Dudley (1924–1930, Fig. 2.6), C. R. Harington (1930–1937, Fig. 2.7) and F. J. W. Roughton (1935–1937, recruited to deal with physiochemical papers), carried on to complete 25 years' service. On his retirement Harden was presented with a silver salver bearing facsimile signatures of those still living who had served with him on Biochemical Society Committees (Fig. 2.8). At the presentation the Chairman of the Committee, Professor H. J. Channon (Fig. 2.9), noted that the number of words published per year in the *Biochemical Journal* when Harden began as Editor was 180,000; this had risen to 1,500,000 25 years later. Harden had thus read around 18 million words in proof, many of them travelling by train to the Lister Institute from his home in Bourne End, near Henley. Hopkins emphasized the importance of Harden's accession to the Editorial chair by reporting that he had had his copy of volume 7 (the first edited by Harden) bound in a special colour.

Fig. 2.7. Sir Charles Harington, F.R.S. Honorary Secretary, 1929–1930. Editor of the *Biochemical Journal*, 1930–1942. Chairman of the Society Committee, 1955–1957. Honorary Member, 1960.

Fig. 2.8. Silver salver presented to Sir Arthur Harden, F.R.S., on 11 March 1938 to mark the occasion of his retirement after 25 years service as Editor of the *Biochemical Journal*.

Fig. 2.9. Professor H. J. Channon, C.M.G. Johnston Professor of Biochemistry, University of Liverpool 1932–1944. Chairman of the Society Committee, 1937–1938.

Harden was such a key figure in the Society in the period between the two World Wars that a short biography outlining his career outside the *Biochemical Journal* is appropriate here. He was born in Manchester on 12 October 1865, the son of a Manchester business man. He was brought up in an austere non-conformist atmosphere and was educated at Tettenhall College in Staffordshire. In January 1881, he entered Owen's College, Manchester to study chemistry under Professor Roscoe, and in 1885 he graduated in the Victoria University with first class honours in chemistry. A year later he was awarded the Dalton scholarship. Then he proceeded to Erlangen and, under the direction of Otto Fischer, prepared α-nitrosonaphthylamine and investigated its properties. Here he was awarded the degree of Ph.D., after which he returned to Manchester, firstly as junior, and later as senior lecturer in chemistry under Professor H. B. Dixon. Harden remained at Manchester for another nine years, during which his activity seems to have been devoted chiefly to teaching and literary work. The literary work to which he was expected to give most of his time was collecting data for the *Treatise on Organic Chemistry* in three volumes by Roscoe and Schorlemmer. According to Chibnall [4] "The same fate awaited Johnny Russell (later Sir John Russell and Director of Rothamsted Experimental Station) … Harden told me that the only remuneration he and Johnny Russell received for their gruelling labours was a brief word of thanks in the third organic volume".

In 1897 he was appointed as chemist to the Lister (then called the Jenner) Institute of Preventive Medicine in London. He had a wide knowledge of chemistry and had proved himself to be a successful teacher and became responsible for teaching the chemical course, which was mostly concerned with the analysis of waters and foods, to medical practitioners desiring to become Medical Officers of Health. These courses were later superseded by special teaching for a Diploma in Public Health conducted in London medical schools and Harden then found that he could devote himself fully to research. At the time Harden was in charge of the Chemical Department at the Institute, but in 1905 it was fused with the Biochemical Department and Harden was placed in charge of the composite department. In 1912, in recognition of his outstanding work on bacterial chemistry and alcoholic fermentation, he was made Professor of Biochemistry in the University of London.

It was during his earliest days at the Lister Institute that Harden started an investigation of the fermentation of sugars by bacteria. Subsequently he embarked on some ten years of research on alcoholic fermentation leading to the discovery of co-zymase and the essential role of phosphoric esters in fermentation. Soon after these discoveries other workers found that phosphorylation provided the clue to many other biological phenomena, including the chemistry of muscle and bone.

During the First World War Harden was left in charge of the Lister Institute as Acting Director and since he wished to devote himself to a subject which would contribute to the War effort, he abandoned his researches on alcoholic fermentation and investigated instead two of the then known accessory food factors or vitamins, lack of which there was good reason to believe was responsible for the diseases beri-beri and scurvy respectively. Both diseases had occurred amongst troops in outposts in Africa and Asia.

Recognition of the importance of Harden's researches came from many quarters. In 1907 he was elected Fellow of The Royal Society, on the council of which he served from 1921 to 1923. In 1929 he shared the Nobel prize for chemistry with von Euler. The Universities of Manchester, Liverpool and Athens conferred honorary degrees upon him and the Kaiserlich Leopold Deutsche Akadamie der Naturforschung of Halle elected him to its membership.

Harden retired from the Lister Institute in 1930 and in the following year he was appointed to its governing body, on which he served until his death in 1940. He became Emeritus Professor of Biochemistry at the University of London in 1931 and The Royal Society awarded him its Davy Medal in 1935. In 1936 he received the honour of a knighthood.

Sir Arthur was elected to the Honorary Membership of the Biochemical Society in 1938. Before his death he willed part of his estate to the Society: the income therefrom was to be applied in defraying the cost of publication of the results of original research in Biochemistry. The emergence of the Harden Conferences catalysed by this bequest is described in the next chapter.

As a person Harden was "somewhat shy and not given to gossip, he disliked public speaking. I recollect the occasion, it was in 1928, when the Society gave a dinner to him and Gowland Hopkins in celebration of their Nobel prizes. Unlike that of the latter, his speech was short, the gist of it being that the prize is given to you for having an idea that worked." [4].

The retirement of Harden clearly marked the end of an era in the history of the *Biochemical Journal*. The overall development of the *Journal* is so crucial to the Society that it deserves a chapter to itself (Chapter 6).

2.3 Emergence of the Biochemical Society

Following the purchase of the *Biochemical Journal*, announced at a special general meeting on 12 October 1912, the Biochemical Club was now poised to transform itself into a Society. This it did at the second A.G.M. on 13 March 1913, when the Committee appointed Gowland Hopkins as its first Chairman. In keeping with the general informality of the Society's organization, it decided to elect a Chairman annually, rather than a President, although in dealing with outside bodies the Chairman would assume the standing of President. This was reaffirmed in 1921. The Society treated "with scorn the Chartered Institution like the Chemical Society with its mace and time honoured formalities" [4]. Plimmer continued as Honorary Secretary and J. A. Gardner was appointed Honorary Treasurer. Plimmer's term of office lasted until 1919, when he moved to the newly founded Rowett Research Institute at Aberdeen. In 1927 it was decided to appoint two Honorary Secretaries: one to deal with Committee business and one to act as a Meetings Secretary. This general arrangement still stands to this day.

J. A. Gardner served with great distinction for 31 years until 1944 and it was not until 1941 that he felt he needed an Assistant Treasurer. The Society was particularly lucky to have Gardner as Treasurer because he could, and did, call on the expert professional help of his brother, T. E. Gardner, of the Chartered Accountant firm of T. Gardner & Son.

The debt which the Society owes to its founder Honorary Treasurer and Honorary Secretary cannot be overestimated but, as will be obvious as this History proceeds, it has always been extremely lucky and/or perspicacious in attracting able persons as Officers. The names of those who helped the Society as Chairmen and Honorary Secretaries from 1911 to 1944 are recorded in Tables 2.5 and 2.6 respectively. Gardner

Table 2.5 Chairmen of the Biochemical Society, 1914–1944

1913-14	F. G. Hopkins, F.R.S.	1929-30	T. S. Hele
1914-15	W. M. Bayliss, F.R.S.	1930-31	T. A. Henry
1915-16	V. H. Blackman, F.R.S.	1931-32	E. Hatschek
1916-17	G. Barger, F.R.S.	1932-33	C. G. L. Wolf
1917-18	A. Harden, F.R.S.	1933-34	R. Robison
1918-19	B. Dyer	1934-35	F. L. Pyman
1919-20	W. M. Bayliss, F.R.S.	1935-36	H. J. Page
1920-21	P. Haas	1936-37	P. Haas
1921-22	S. B. Schryver	1937-38	H. J. Channon
1922-23	R. H. A. Plimmer	1938-39	R. A. Peters, F.R.S.
1923-24	J. C. Drummond, F.R.S.	1939-40	R. H. A. Plimmer
1924-25	P. Hartley	1940-41	G. M. Findlay
1925-26	H. W. Dudley, F.R.S.	1941-42	D. P. Cuthbertson
1926-27	C. Lovatt Evans, F.R.S.	1942-43	J. C. Drummond, F.R.S.
1927-28	Ida Smedley-Maclean	1943-44	J. V. Eyre
1928-29	R. A. Peters, F.R.S.	1944-45	E. C. Dodds, F.R.S.

Table 2.6. Honorary Secretaries of the Biochemical Society, 1911–1945

1911-19	R. H. A. Plimmer
1919-22	J. C. Drummond, F.R.S.
1922-24	H. W. Dudley, F.R.S.
1924-27	P. Hartley
1927-29	H. D. Kay, F.R.S. & R. Robison
1929-30	R. Robison & C. R. Harington, F.R.S.
1930-38	A. C. Chibnall, F.R.S. & H. Raistrick, F.R.S.
1938-40	A. C. Chibnall, F.R.S. & F. G. Young, F.R.S.
1940-43	F. G. Young, F.R.S. & W. T. J. Morgan, F.R.S.
1943-45	W. T. J. Morgan, F.R.S. & W. Robson

was the only Honorary Treasurer during this time. The longest serving Secretary during this period was A. C. Chibnall, eventually Sir William Dunn Professor of Biochemistry in the University of Cambridge (Fig. 2.10; see also Plate 1B). In an autobiographical essay in 1966 [5] he painted a clear picture of the informality and bonhomie which prevailed in the Society right up to 1940:

"The only outside commitment I had in those days (1929), was that of Committee Secretary to the Biochemical Society, H. Raistrick dealing in a similar capacity with the business connected with the public meetings. We ran together in harmony with the Treasurer J. A. Gardner, for seven years, foregathering one afternoon each year to check the books and to dine later with Gardner as host. Although the Society was flourishing and its membership had passed the seven hundred mark, the Journal was eating up all our available cash, and as secretaries, our official attendances at meetings, even as far away as Aberdeen, had to be at our own expense. Raistrick and I between us knew almost every member except those few who lived abroad, and the Society to us was just a happy family with Harden and Harington shouldering all the burden of publication."

Fig. 2.10. Professor A. C. Chibnall, F.R.S. Sir William Dunn Professor of Biochemistry, Cambridge 1943-1949. Honorary Secretary, 1930-1940. Honorary Member, 1969.

In 1986, as a very young 93-year-old, Chibnall still remembered those times vividly and realized that the end of an era was rapidly approaching for the Society: "Towards the end of my period of office the number of our members employed in industry was on the increase, and partly because of this I think the subject was beginning to fragment and discussion at meetings was becoming less breezy and spontaneous. In retrospect the seeds of the formal institute were beginning to sprout." [4].

Formalization of appointment of Officers did not take place until 1943, when it was decided that the terms of Office should be seven years.

2.4 Financial Position of the Society

It is now generally accepted that the income from the sale of the *Biochemical Journal* to non-members of the Society is the major source of revenue for the Society. It has not always been so. The annual subscription rate for members was set at £1 5s [about £30 today], which included provision of the *Journal*. However, costs of publication rose after the First World War and the Society was kept solvent after 1925 by the use of accumulated profits, occasional gifts, a grant from The Royal Society, rare payment by authors of part or all the costs of printing long papers, a more generous contract from the Cambridge University Press and an increase in subscription rate to £1 15s (£1.75) [£32]. In addition outside subscribers were charged more. By 1931 losses once again appeared on the balance sheets but a further increase in subscription rate, to £2 2s (£2.10) [£43], another grant from The Royal Society, minor concessions from the Cambridge University Press, together with a new outside subscription rate of £3 10s (£3.50) [£72], allowed a balance to be struck.

The hazardous financial position of the Society at this time has been amusingly described by Chibnall [4]:

> "Our Finance during my period of office was always in a precarious state, and when Gardner's brother, our accountant, had given us our statement for the year, Gardner and I used to visit a representative of the Cambridge University Press at its warehouse in Euston Road, London, to discuss payment of our bill for the printing and distribution of the Journal. Our discussion was always quite amicable and to the best of my recollection we left with the payment of our bill still two to three years in arrears! As the Press representative used to tell us with a smile, we were supported from profits on the sale of the Bible which it was entitled to print under a charter of James I."

It was against this early background that, in May 1925, T. Gardner & Co. strongly advised the Society that it should become incorporated. The Committee felt no sense of urgency over this and it was not until 1928 that they formed a subcommittee to examine the options open to the Society. The subcommittee rejected the idea of a Royal Charter and of the formation of a limited liability company; the remaining options were (i) incorporation under Section 20 of the Companies Act, which was designed by the Board of Trade to meet organizations, such as learned societies, which wished to have the status of a learned body without being styled 'Limited', or presumably today 'plc', and (ii) a Trustee system. The subcommittee failed to make a clear recommendation and later in the year the full Committee decided, by a narrow majority, in favour of a Trustee system. The original Trustees were J. L. Baker, H. W. Dudley, J. A. Gardner, A. Harden, H. D. Kay and R. H. A. Plimmer. By this time, however, measures just outlined, accompanied by economies in printing (smaller type and a two column format) and coinciding with a rapid increase in membership and in the number of outside subscriptions, resulted in some profit which could be invested for future developments. The matter was not raised again until 1944 when confidence in the Trustee system was affirmed. However, as the Society grew rapidly after the Second World War, reappraisal became urgent and the events leading to the decision to become incorporated in the early 'sixties are described in Chapter 3.

2.5 General Developments

It was agreed when the Society was founded that eight meetings a year would be held and this was generally adhered to during 1911–1944. Up to the early 1920s the attendance at meetings averaged between 40 and 50 but by the 1940s it occasionally reached over 400. A fascinating side-light on the early days comes from Chibnall [4]: "Very noticeable in those days was the virtual absence of members from Cambridge. This was because in Hopkins' laboratory the teaching was 'how it works', based on physiology, whereas elsewhere it was 'what is it', based on (medical) chemistry". In the 'thirties, efforts were made to organize some sort of Scottish Association to arrange meetings in Scottish centres, a powerful argument being the lack of travel funds for visits to London and other English Centres. However, the Committee eventually agreed that at least one meeting a year should be held in Scotland, which was an acceptable compromise. The number of members of the Society steadily increased (Fig. 2.11) from 132 at the first A.G.M. in 1912 and it reached the 1000 mark on 1 January 1944. The membership now (1986) stands at around 6500 (see Chapter 3).

Landmarks which were reached were the hundredth meeting on 13 March 1926 and the 21st Birthday Anniversary Meeting at UCL on 17 November 1933. At the first meeting, a collection of signatures of those attending the celebration

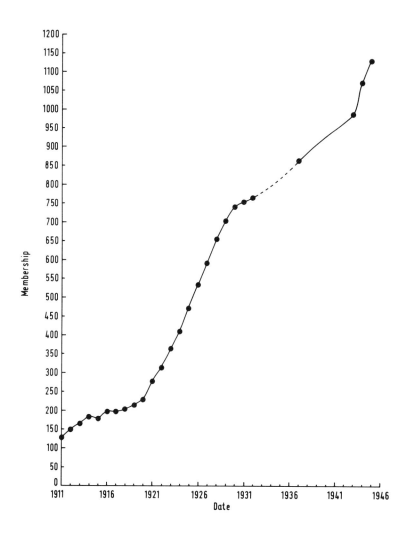

Fig. 2.11. Numbers of members of the Biochemical Society from 1911 to 1944.

dinner at the Grosvenor Hotel was framed and presented to the Lister Institute for custody. It now hangs in the Society's headquarters at Warwick Court (Fig. 2.12). At the second meeting a historic photograph was taken of those original members of the Society attending the meeting (Fig. 2.13).

A special dinner at the Hotel Victoria, London, was held on 3 February 1930 to celebrate the award of the Nobel Prize for 1929 to three distinguished members of the Society, F. G. Hopkins (Medicine) and H. von Euler and A. Harden (Chemistry) (Fig. 2.14). A number of other members also won the Prize during the 1911–1945 period, the first being S. A. S. Krogh in 1920. A full list of Nobel Laureates of the Society between 1911 and 1945 is given in Table 2.7.

2.6 Honorary Members

The highest accolade which the Society can bestow on members is Honorary Membership. At this time (1911–1944)

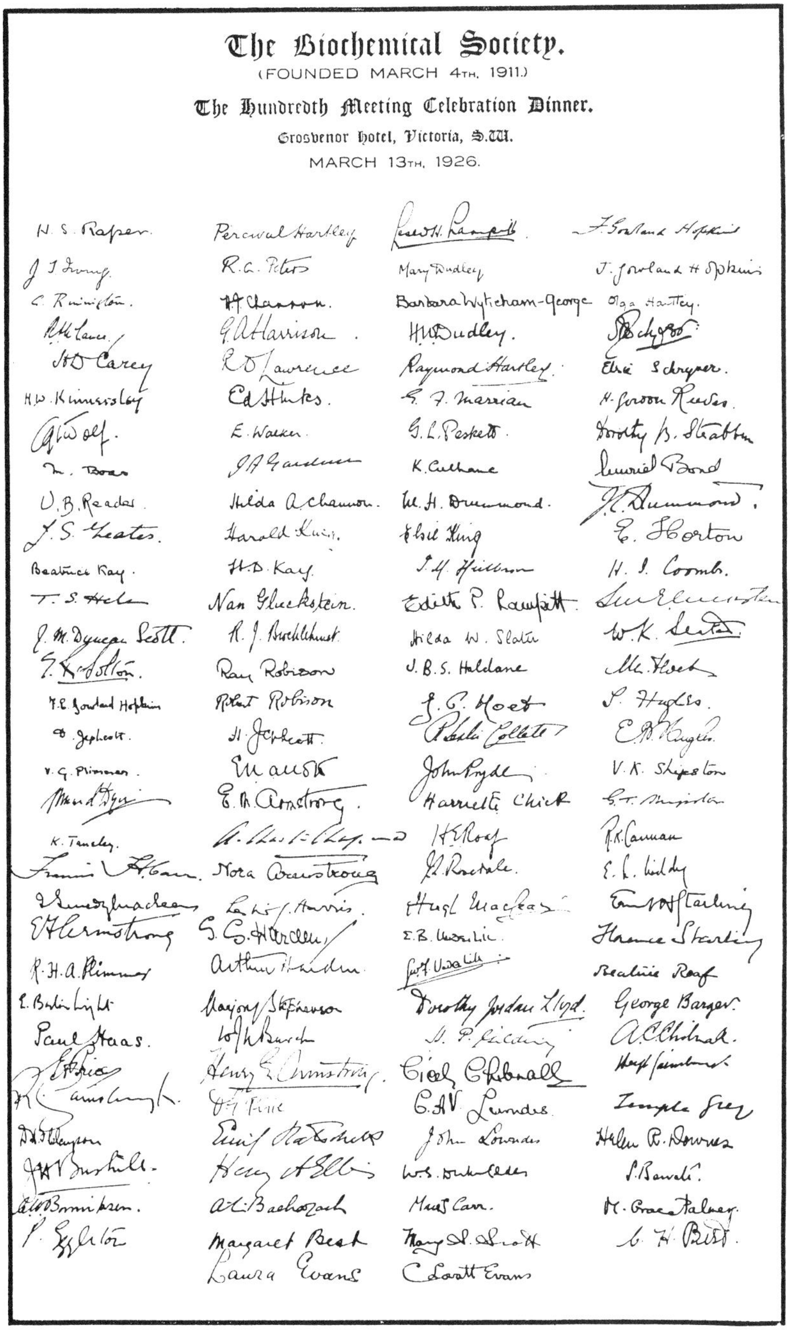

The Biochemical Society.
(FOUNDED MARCH 4TH. 1911.)
The Hundredth Meeting Celebration Dinner.
Grosvenor Hotel, Victoria, S.W.
MARCH 13TH. 1926.

Fig. 2.12. Signatures of members present at the dinner celebrating the hundredth meeting of the Society, 13 March 1926.

apart from the major criteria of high academic distinction and exceptional service to the Society the only rules for election to Honorary Membership were that no appointment should be made whilst a member held an official appointment with the Society and was actively engaged in research (i.e. not retired). However, on election the Honorary Members were deprived of the right to vote. The election of W. D. Halliburton in 1923

Fig. 2.13. Group of original members present at the 21st Anniversary dinner at the Grosvenor Hotel, 17 November 1933. Left to right: (back row) F. P. Worley, C. Lovatt Evans, P. Hartley, E. L. Kennaway, H. W. Bywaters, W. H. Hurtley; (middle row) J. V. Eyre, S. A. Mann, C. Dorée, J. K. Close, H. J. Page, S. G. Paine, J. Golding, W. Cramer; (front row, seated) E. Mellanby, W. Ramsden, R. H. Plimmer, J. A. Gardner, F. G. Hopkins, A. Harden, C. J. Martin, B. Dyer, T. A. Henry.

as the first Honorary Member has already been noted (Chapter 1) and contemporary members must have been delighted to hear of the election in 1930 of A. Harden and F. G. Hopkins, and of R. H. A. Plimmer in 1943.

2.7 Discussion Meetings

The organizing of meetings for discussion was proposed at the second meeting of the Club in 1911 and the first was held in December 1914, when "Micromethods of Analysis" were demonstrated at the Lister Institute. Another, which was a joint meeting with the Society of Public Analysts, was held in May 1915. This was the first joint meeting held by the Society and the subject discussed was "Methods Adopted for the Estimation of the Nitrogenous Constituents of Extracts from Albuminous Substances, such as Meat Extracts, with Special References to the Interpretation of the Results". At least no-one present could complain that they were lured to the meeting under false pretences! An early discussion meeting threw up one of the best anecdotes in the Society's history, related by N. Pirie. In October 1924, J. B. S. Haldane, then Reader in Biochemistry at Cambridge, and colleagues discussed the effect of inducing acidosis in Haldane (Haldane was never unwilling to use himself in experiments). After a longish silence for discussion, Sir Charles Martin commented: "these are very interesting and important findings, so interesting and important indeed that they ought to be repeated on a normal subject" [3].

Fig. 2.14. Frontispiece of the menu for dinner in honour of the Society's Nobel Laureates in 1929.

Other Societies with which joint meetings were held in the early days were the Physiological Society (1918), the Society of Chemical Industry (London Section, 1923, 1926, 1927) and the Institute of Brewing (1923). Later the Pathological Society (1944) and the Nutrition Society (1944) were similarly involved. As will be obvious in later chapters, the policy of joint meetings has continued with only occasional gaps and the constant willingness to take the initiative and to discuss key

Table 2.7. Members of the Biochemical Society who were awarded Nobel Prizes, 1911–1942

Details for 1945 on are given in Table 3.17.

	Chemistry*		Medicine
1927	Prof. H. O. Wieland	1920	S. A. S. Krogh
1929	Sir Arthur Harden, F.R.S.	1923	A. V. Hill, F.R.S.
	H. K. A. S. von Euler		J. J. R. Macleod
1937	Sir Norman Haworth, F.R.S.	1929	Sir F. Gowland Hopkins, O.M., F.R.S.
	Prof. P. Karrer, Foreign Mem. R.S.		
1939	Prof. A. F. J. Butenandt, Foreign Mem. R.S.	1936	Sir Henry Dale, F.R.S.
			O. Loewi
	Prof. L. Ruzicka, Foreign Mem. R.S.	1937	A. Szent-Gyorgi

*T. Svedberg (1926) was not a member of the Society at the time of his award.

areas of biological research with other Societies has been a great strength of the Society. It has emphasized the integral position of Biochemistry in modern biology and helped to prevent various specialized aspects being hived off as separate societies.

The biggest gap in the formal organization of discussion meetings occurred between 1928 and 1934, when the matter was raised in Committee by H. J. Channon. It was probably stimulated by the knowledge that the Chemical Society was active in the field and the Committee decided to approach the Chemical Society unofficially about the possibility of taking part in their discussion meetings. The approach bore fruit and within three months it was reported that the "Biochemical Society would be officially invited to help arrange and take part in biochemical discussions which the Chemical Society would hold". The first such meeting was that on "The Chemistry and Biochemistry of Lipoids" proposed for later in 1934. Furthermore in 1937 the Committee decided to hold one discussion meeting a year additional to the normal eight meetings of the Society. In 1940 a joint discussion meeting was held between the Faraday, Physiological, Biochemical and Chemical Societies on "Chemical Structure in Relation to Membrane Permeability".

In 1942 the Committee acknowledged the widespread desire amongst the members for discussion meetings and also agreed that a 700 word summary of the main papers presented at these meetings should be precirculated. Inevitably, after this slight opening of the stable door, there was a request that the proceedings of the next discussion meeting on "Tetrapyrrolic Pigments" should be published *in toto*. This was not accepted by the Committee but as we shall see in Chapter 3 the pressure for publication mounted and soon a very successful Symposium series was established.

In May 1944 the Committee further acknowledged the increasing importance of discussion meetings in the Society's activities by accepting the recommendations of a special subcommittee set up to look into the proposal that at least two discussion meetings be held per year and that they should be an integral part of the meetings programme. The main Committee did not accept a further rather odd proposal that 'for the time being' such meetings should be held only in Oxford, Cambridge or London.

2.8 Proceedings

In the early days there was no outlet for permanently recording the proceedings of the meetings of the Society but in 1924 the Editor (S. Miall) of the newly established journal, *Chemistry and Industry*, offered the Society the hospitality of his pages for prompt publication of short abstracts of papers presented at the Society's meeting. This proposal appealed to the Committee and was gratefully accepted. The practice was continued until 1941, when the possibility of printing unrefereed abstracts in the *Biochemical Journal*, rejected by the Committee in 1926, was reopened. The new proposal was accepted with the provisos: (i) that the abstracts were printed in a style different from that of the *Biochemical Journal* and (ii) that a clear statement was made absolving the Editors (of the *Biochemical Journal*) from responsibility for the content and method of presentation of the abstracts.

References

1. Plimmer, R. H. A. (1949) *The History of the Biochemical Society 1911–1949*. pp. 24. The Biochemical Society, London.
2. Lovatt Evans, C. A. Manuscript deposited in the Biochemical Society's Archives. Date unknown (? 1930s).
3. Morton, R. A. (1969) *The Biochemical Society: its History and Activities 1911–1969*. pp. 160. The Biochemical Society, London.
4. Chibnall, A. C. (1986) Unpublished autobiography [appropriate section kindly made available to the author; now available from the Biochemical Society (price £2.00)].
5. Chibnall, A. C. (1966) The road to Cambridge. *Annu. Rev. Biochem.*, **35**, 1–22.

Chapter 3

General Developments 1944–1986

3.1   Introduction
3.2   Administrative Developments
3.3   Legal Status of the Society
3.4   General Post-War Planning
3.5   Biological Council
3.6   Anniversary Meetings
3.7   General Pattern of Ordinary Meetings
3.8   Proceedings, Agenda Papers, *Bulletin*
3.9   Travel Funds
3.10 Medals and Named Lectures
3.11 Fellowships and Scholarships
3.12 Awards Committees
3.13 The Harden Conferences
3.14 Honorary Membership
3.15 The Society's Nobel Laureates
3.16 A Royal Charter — To Be or Not To Be?
3.17 The Chemical Society Library
3.18 Archives and the Science Museum
3.19 The Society's Logo

3.1   Introduction

As with many aspects of our national life, the years 1944–1985 can be considered a watershed in the development of the Biochemical Society. The end of the Second World War left the Country exhausted but a spirit of optimism was in the air. Thanks to the efforts of the Honorary Officers, the Society successfully survived the War and the mood of optimism within the Society was fully justified and has lasted well after the hopes of a brave new world have long faded in other areas of human activity. Biochemistry rapidly developed into a thriving discipline and this has been maintained throughout the post-War period, although in recent years the pace has slackened somewhat mainly owing to the parsimony of recent Government policy on support of Science.

This blossoming of Biochemistry in the post-War years has been one of the great scientific successes of all time and in the

U.K. the simultaneous expansion of the Biochemical Society has been equally impressive. The number of members has increased smoothly for nearly 30 years, from 1017 in 1946 to 5877 in 1973, from when there was an overall slight downward trend until 1981. Since then there has been a very marked increase (Fig. 3.1) and in early 1986 the membership topped the 6500 mark. A more detailed study of the changes in the membership over the past 10 years (Table 3.1) shows

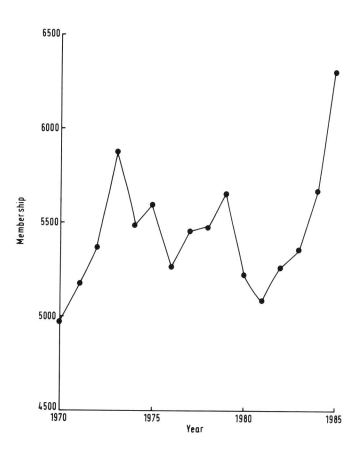

Fig. 3.1. Changes in membership of the Society between 1970 and 1985.

Table 3.1. Changes in the pattern of membership of the Biochemical Society, 1976–1985

	1976	1977	1978	1979	1980	1981	1982	1983	1984	1985
TOTAL B/F	5186	5263	5447	5474	5654	5221	5082	5258	5356	5662
New members	381	513	456	587	434	390	463	487	638	1111
(includes students)	(167)	(242)	(238)	(296)	(223)	(209)	(277)	(253)	(271)	(520)
Resigned	89	61	85	35	144	136	142	47	25	96
Deceased	17	8	12	25	13	16	17	19	5	35
Lapsed	365	260	332	347	432	377	128	323	302	337
Total C/F	5263	5447	5474	5654	5221	5082	5258	5356	5662	6305

that new student membership fluctuates around 50% of the total membership. Emeritus membership, a new category established in 1971 with 98 members, now stands at 356. Overseas members come from 60 countries and a recent zonal breakdown is given in Table 3.2. In 1985 the highest number of members in Western Europe was from Spain (90), and the largest number (795) in the 'overseas' category was inevitably found in the U.S.A. A total of 31 members in South America, 12 of which are in Argentina, is surprisingly low. The one member in the U.S.S.R. must feel very lonely.

The post-War increase in membership alone demanded organizational changes but the additional commitments which the Society was also shouldering, some inevitable and some innovative, soon made these changes an urgent necessity. By the early 1960s the Society could no longer be run by Honorary Officers working on sufferance in University offices on a shoe-string budget. A headquarters with professional staff was clearly needed and this was eventually achieved in 1966. Many felt that this development took too long to materialize, but as we shall see in the next section, many problems had to be overcome before the Society could settle down in its headquarters in Warwick Court, which provided a much improved base as well as being an impressive capital investment.

Table 3.2. Geographical distribution of members in 1985

Zone	Number*	
North America	795	(78)
South America	31	(2)
Western Europe (excluding U.K.)	644	(52)
Eastern Europe	20	(1)
Australia	91	(1)
Africa	37	(−1)
Japan	29	(2)
Overseas	135	(23)
United Kingdom (including Ireland)	4503	(479)
Total	6285	(637)

*Numbers in parentheses indicate the change in membership from 1984 to 1985.

3.2 Administrative Developments

Immediately after the Second World War the Honorary Secretaries worked from their University bases and used departmental secretaries when and if they were available. Christmas boxes were solemnly voted by the Committee each year to the ladies concerned.

Fig. 3.2. Professor E. J. King. Chairman of the Editorial Board, 1946-1952. Chairman of the Society Committee, 1957-1959.

With increasing activities both in general administration and in running the *Biochemical Journal* it was inevitable that demands for full time staff and proper office accommodation would arise. The pressure first came from the Chairmen of the Editorial Board of the *Biochemical Journal*, who were faced with a rapid increase in the number of papers being submitted. The first move was made by E. J. King (Fig. 3.2) (Chairman of the Editorial Board, 1946-1952), who ran the journal from his department at Hammersmith, in accommodation rented from the University of London.

When A. Neuberger (Plate 3A) took over from E. J. King, the Editorial Office was moved to the National Institute for Medical Research and the Society paid £50 [£425] p.a. rent for "two years plus telephone". A more permanent solution was required but discussion in 1954 with the Linnean Society to obtain accommodation in their rooms at Burlington House came to nothing. However, living once again from hand to mouth, the Editorial Office moved into temporary quarters (two rooms in the Director's former top-floor flat) in the Lister Institute in 1955. This accommodation became available when it was vacated by Biological and Medical Abstracts Ltd. The rent was £20 [£170] p.a. plus £20 p.a. for library facilities. Attempts to obtain a more permanent arrangement failed, as did approaches to the University of London for office space. The inevitable ultimatum came in October 1959 from the Director of the Lister Institute, who required the space by Christmas 1959 or March 1960 at the latest. The Committee hoped to take up accommodation in the new Medical Research Council (MRC) premises in Park Crescent but this was not due to become available until 1961. Temporary offices were eventually found in 133-135 Oxford Street at a rent of £675 p.a. [£4750], excluding rates, on a 7 years' lease. After strong pressure by Harington in the face of Treasury obstructions an agreement with the MRC for the use of office accommodation was concluded in February 1961 at a total rent of £1425 [£10,000] p.a. and the Editorial Office, now comprising three full time and two part time staff, moved there in September 1961. By December 1961 the Oxford Street offices had been sub-let on favourable terms. Apparently the MRC was happy to have a Learned Society temporarily in four rooms in the basement of its premises at Park Crescent as it made their negotiations with the Treasury somewhat easier.

The first official move in the effort to obtain general office, rather than Editorial Office, accommodation was in 1950 when the Committee Secretary represented the Society on a Scientific Societies Accommodation Committee set up by The Royal Society to consider the proposals initiated by our Secretary for a new Science Centre. However, it was noted that the Centre was not expected to be completed for 10 years. In fact, the project never materialized, and it was not until 1960 when P. N. Campbell (Plate 1B) was elected Honorary Secretary that he persuaded the Committee to provide him with a part-time secretary.

In 1961 the Society appointed an Administrative Secretary, Mr G. McHardy, and he was housed temporarily in UCL from April 1961 until he joined up with the Editorial Administration at Park Crescent.

The breathing space provided by the tenancy at Park Crescent allowed the Committee to look to the future. A Finance Sub-Committee to advise the Treasurer on investments and general financial policy was set up. It recommended that the Society should have its own headquarters. After a number of abortive efforts, 7 Warwick Court (Fig. 3.3) was bought in 1966 for £57,094 [£345,000]. It contained a small flat which was sub-let. This was by far the largest financial transaction which the Society had undertaken and was made possible only by selling about 70% of the Society's investment. The responsibility for this decision was a heavy one on the Officers concerned; but the future has more than justified their action and the Society must be forever grateful that they took the decision they did.

At the time the purchase of Warwick Court was being negotiated, the Administrative Secretary resigned and a new post, Executive Secretary, was established. Mr A. I. P. Henton (Fig. 3.4) was appointed and by 1969 the senior office staff numbered nine.

In the early 'seventies there was a feeling of optimistic expansionism in the air in the U.K. and nowhere was it more apparent than in Biochemistry. The lively Honorary Officers and most Committee members responded to this general feeling and wished to develop the activities of the Society, particularly in publications. These views were not without their opponents and many doughty battles were fought in Committee, some of which are recalled later in this chapter. However, the 'progressives' eventually won but implementation of their plans was obviously quickly going to overstretch the facilities at Warwick Court. Perhaps even more importantly the basic activities associated with current commitments were also increasing rapidly. So, only after a few years Warwick Court became too small to handle expansion on two fronts and the need for larger premises became acute.

Wide searches in London revealed no suitable accommodation available at a price which was not wildly beyond the Society's means. Removal of the office to the provinces had financial attractions but the disruption of the domestic arrangements of the permanent staff was not acceptable. Advice was sought from the Location of Offices Bureau, an organization set up to help those seeking accommodation outside London. Eventually warehouse and office accommodation, recently built and unoccupied, was found on the

Fig. 3.3. 7 Warwick Court, London WC1R 5DP. Headquarters of the Society since 1966.

Fig. 3.4. A. I. P. Henton. Executive Secretary, 1966–1984.

Fig. 3.5. The Society's Depot at Colchester, Essex.

Whitehall Industrial Estate just outside Colchester (Fig. 3.5). However, the building, which was for sale on a long lease, was too big for the Society's needs but it was possible to divide it into two self-contained units, one of which could be let. Dr D. F. Elliott (Plate 4A), the Honorary Treasurer who masterminded these activities, writes:

> "At the price being asked, purchase was an attractive proposition and it was decided to proceed. We were indeed able to find occupants for one half of the building at a rental which partly offset the loss of income resulting from the sale of the investments required to raise the capital for the project, the investments chosen having made substantial capital gains since they were acquired. Some expenditure was also needed for the modernization of equipment moved from Warwick Court and for the fitting out of the building as a book depot. With the appointment of a Publication Manager (Mr A. Sabner: Fig. 3.6) and an assistant from amongst our senior staff, the Colchester enterprise commenced operations on 12 June 1972. There was now ample room for growth of the *Journal* and for any additional publishing projects that were likely to occur. It was also intended that income would be raised by taking on distribution work for sister societies lacking such facilities themselves."

When the publishing activities other than the Editorial Office were moved out of Warwick Court, it was possible to refurbish the basement to provide an attractive Committee

Fig. 3.6. A. S. Sabner. Manager of Colchester Depot, 1972–1985.

Fig. 3.7. The refurbished Committee Room at 7 Warwick Court.

room (Fig. 3.7). During these alterations a major structural fault was disclosed which had to be dealt with immediately. The rest of the building was also renovated in order to comply with the Greater London Council fire regulations — total cost over £30,000 [£120,000].

It is now over 10 years since the additions to and upgrading of the Society's premises, and up to 1985 they proved adequate for all the diversification which had taken place. However, many new projects are scheduled for the future and once again larger premises are required. By the time this appears in print new accommodation in Central London may well have been bought.

In 1984, Mr G. D. Jones (Plate 2A) took over the post of Executive Secretary from Mr Henton who, in his 18 years with the Society, had had to deal with what is likely to emerge on a relative scale as the most expansionist era of the Society. The office is now being developed according to new organizational plans summarized in Fig. 3.8. This development will be expanded on later but it should be noted that two new posts have been established in response to recent requirements, an Assistant Meetings Officer and a Research and Information Officer. These important changes have been achieved without the need to increase the overall number of staff employed.

Since a headquarters was first established in the 1960s, the Society has generally been extremely fortunate in its choice of professional staff who have exemplified a true spirit of co-operation and loyalty, none more so than the present senior staff: Doris Herriott, Meetings Office (Plate 2A) and Tony Evans, Editorial Manager (Plate 2A).

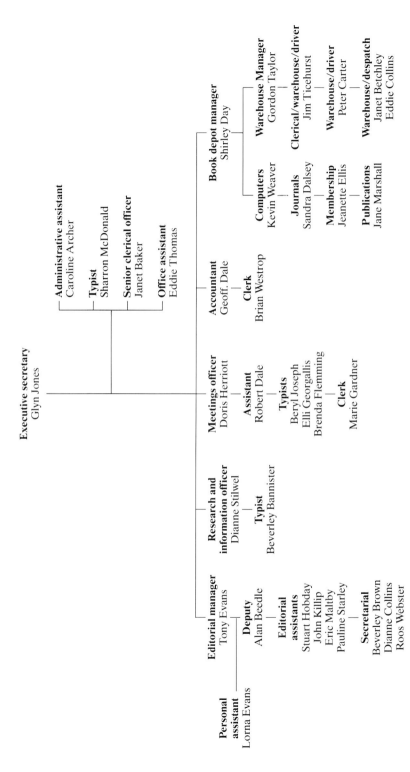

Fig. 3.8. Administrative organization of the Society (1986).

3.3 Legal Status of the Society

The Trustee system of control of the Society, preferred by members before 1944 (Chapter 2) was still favoured after the Second World War and in 1954 a supplementary Trust Deed was sealed which vested the Society's property in the Trustees. In 1960 the Committee made new rules regarding the function of the Trustees and in 1962 agreed that minutes of Committee meetings should be sent to the Convenor of the Trustees (Dr J. H. Bushill, Honorary Treasurer, 1944–1952, see Chapter 4) to keep them "better informed of the Committee's decisions and activities", but at their next meeting they decided to send the minutes to all Trustees.

However, the position of the Trustees had become equivocal after the change of rules in 1960 and the problem came to a head in 1964 when Sir Charles Harington, Chairman of the Trustees, voiced the Trustees' disquiet: "The Trustees were now clearly bound to act on the instructions of the Committee and felt some uneasiness in that they were accepting a measure of responsibility for funds over which they had no control in respect either of expenditure or investment policy". As the Trustees did not see a resolution of this problem within the constitution of the Society, they suggested that the Committee might consider incorporation under the Companies Act. The Committee accepted this suggestion, which it will be remembered (Chapter 2) had been first mooted in 1928. Their deliberations were catalysed by the knowledge that the Society's activities were now 'big business' with a revenue account of around £100,000 [£700,000] and that, in an inflationary period, capital funds tied up in Trustee stocks were not the best use of its money. By September 1964, plans for incorporation had been approved by the Committee and it was agreed to discuss them at the A.G.M. in 1965. However, arrangements were speeded up when early in 1965 the Society received Counsel's opinion that under the present rules payment of Honoraria to editors *et al.* was illegal. Documents and explanatory notes were rapidly prepared and circulated to members so that a decision could be made at a Special General Meeting called in Oxford on 15 July 1965. The proposal was approved after a long and detailed discussion and on 25 November 1965 the Biochemical Society became incorporated under the Companies Act. The Board of Trade agreed that the word 'Limited' could be omitted from the title. However, as members will have observed, official Society notepaper now carries the statement "A Company Limited by guarantee, registered in London No. 892796".

The required "Memorandum and Articles of Association of the Biochemical Society" were drawn up, referring to a company limited by guarantee but not having a share capital. The objectives of the Society and its methods of running its financial affairs are detailed in the Memorandum. The Articles were drawn up after long discussions; they contain nineteen sections dealing with membership and subscriptions, twenty-one with organization, five with general meetings, seven with vote of members, one with the seal, three with publications, four with accounts, two with audit, two with notices, one with dissolution and one with indemnity to officials. With only one or two minor instances these Articles have worked well and have in no way hamstrung the Committee in its efforts to improve and expand the Society's activity. An early embarrassment, however, was to discover that the Committee had no power to co-opt members. This was put right at an A.G.M. in Liverpool in 1968, when it was agreed that both the Symposium Organizer and the Secretary-General of FEBS (while he is a British member of the Biochemical Society) could be co-opted. Minor changes have been made periodically, and after 20 years the Articles probably need refurbishing, especially in relation to EC rules; this is likely to happen in the near future.

On the advice of the then Honorary Treasurer (W. F. J. Cuthbertson, see Chapter 4), a Finance Sub-Committee was set up in 1966, comprising the Honorary Treasurer (Chairman), the Honorary Committee Secretary, a representative of the Editorial Board of the *Biochemical Journal* and three other Committee members. The Sub-Committee was consolidated into the Finance Board in 1973.

The success of the incorporation of the Society was largely due to the skills of the Honorary Treasurer, the Executive Secretary (A. I. P. Henton) and the Chairman of the Society at that time, Professor Helen Porter (Fig. 3.9).

With incorporation the role of the Trustees came to an end. They had guided the Society through various financial problems for almost 40 years; they acted with great acumen and laid the foundations for the financial prosperity the Society enjoys today. A perusal of the list of Trustees who have helped the Society over the years (Table 3.3) immediately reveals not only their biochemical eminence but also recalls the many other ways in which they have helped the Society.

Incorporation did not alter fundamentally the basis on which the Society was run, that is by a general Committee and

Fig. 3.9. Professor Helen K. Porter, F.R.S. Chairman of the Society Committee, 1965–1967.

Table 3.3. Trustees of the Biochemical Society

	Dates		Dates
J. L. Baker	1929	Sir Charles Harington, F.R.S.	1929–65
H. W. Dudley, F.R.S.	1929–38	A. C. Chibnall, F.R.S.	1938–65
J. A. Gardner	1929–46	Sir Jack Drummond, F.R.S.	1942–65
Sir Arthur Harden, F.R.S.	1929–42	H. Raistrick, F.R.S.	1946–65
H. D. Kay, F.R.S.	1929–65	J. H. Bushill	1956–65
R. H. A. Plimmer	1929	Sir Rudolph Peters, F.R.S.	1956–65

Table 3.4. Biochemical Society Chairmen, 1945–1986

Dates	
1945–46	A. C. Chibnall, F.R.S.
1946–47	F. A. Robinson, C.B.E.
1947–48	Margaret M. Murray
1948–49	W. T. J. Morgan, C.B.E., F.R.S.
1949–50	H. Raistrick, F.R.S.
1950–51	F. Dickens, F.R.S.
1951–52	Sir Charles Dodds, F.R.S.
1952–53	Sir Rudolph Peters, F.R.S.
1953–55	Sir Frank Young, F.R.S.
1955–57	Sir Charles Harington, F.R.S.
1957–59	E. J. King
1959–61	R. A. Morton, F.R.S.
1961–63	J. N. Davidson, C.B.E., F.R.S.
1963–65	F. C. Happold
1965–67	Helen K. Porter, F.R.S.
1967–69	A. Neuberger, C.B.E., F.R.S.
1969–71	G. A. D. Haslewood
1971–74	T. W. Goodwin, C.B.E., F.R.S.
1974–77	T. S. Work
1977–80	R. R. Porter, C.H., F.R.S.
1980–83	S. V. Perry, F.R.S.
1984–86	K. S. Dodgson
1986–	H. M. Keir

Table 3.5. Honorary Secretaries of the Biochemical Society, 1945–1986*

Dates		Dates	
1945–47	W. Robson	1964–69	K. S. Dodgson
1947–52	J. N. Davidson, C.B.E., F.R.S.	1959–61	W. J. Whelan
1950–53	L. Young	1967–73	A. N. Davison
1952–55	R. H. Thompson, C.B.E., F.R.S.	1973–74	A. P. Mathias
1955–58	F. L. Warren	1973–80	J. B. Lloyd
1955–59	C. E. Dalgliesh	1980–85	D. Robinson
1959–64	P. N. Campbell	1981–	R. H. Burdon
1962–67	H. R. V. Arnstein	1985–	A. D. B. Malcolm

*In many cases Honorary Secretaries started as Meetings Secretaries and then moved on to General Secretaries: hence the frequent overlap of dates.

Honorary Officers. The Committee Chairman and Honorary Secretaries who have served the Society are named in Tables 3.4 and 3.5 respectively. The Honorary Treasurers are referred to in detail in Chapter 4. In 1953 the Chairman's period of office was extended to two years and in 1971 to three years. However, it was not until 1958 that it was agreed to pay the Chairman's expenses at the same rate as that for the Officers. This, it was suggested, would help to make him "less anonymous than he had been in the past". The office of International Secretary was established in 1964, when W. J. Whelan (Fig. 3.10) was appointed, but it was abandoned in 1970 when, owing to the formation of FEBS (see Chapter 7), it became superfluous. A. P. Mathias (Plate 1B) was the only other holder of this office (1968–1970). However, because of the

Fig. 3.10. Professor W. J. Whelan. Honorary Secretary, 1959–1962. Honorary International Secretary, 1964–1967.

increased extra-European activities of the Society in the past few years, there is pressure for the office to be restored. The Chairman of the Advisory Committee for Publications became Secretary for Publications when that Committee was upgraded to the Publications Board in 1973. Similarly the Chairman of the Professional and Education Sub-Committee became an Honorary Officer when this Committee was upgraded to the Professional and Educational Committee in 1985.

3.4 General Post-War Planning

The development of the Biochemical Society has depended to a great extent not on detailed formal planning but to informed opportunism and outstanding scientific entrepreneurism by a group of young, talented and enthusiastic Officers who often laboured long into the night (sometimes until 4 a.m., with appropriate liquid sustenance) on behalf of the Society. They had the knack of responding to the pressure of events whilst, at the same time, initiating many of the events. It is to their great credit that they rarely put a foot wrong although in their enthusiasm they sometimes offended the susceptibilities of more cautious and conservative Committee members and they occasionally found the democratic process too ponderous for their needs. An example of the latter is quoted by Morton [5]. "In 1964 exception was taken to the informal way in which a retiring Honorary Secretary tended to find his own successor. The criticism was supported and the Committee decided that in future there should be a nominating Committee consisting of the Chairman, the Honorary Officers and two ordinary members of the main Committee".

Although many developments within the Society, particularly on the international scene (Chapter 7), took place between 1944 and 1965, it was not until the latter date that the Committee discussed the future of the Society in detail and set up a planning Sub-Committee. This reported in the middle of 1966 and indicated that a major expansion should be in publications. This view was based not only on the financial worry that the Society had only one source of income, the *Biochemical Journal*, but also to the fear partly because of this that the *Biochemical Journal* was losing its appeal, because of its breadth of coverage in the rapidly expanding science of Biochemistry and was not attracting the most exciting research papers. The Committee accepted that the formation of Groups, which had recently begun to emerge spontaneously, should be encouraged and that Group meetings should be held at the same time as the ordinary meetings of the Society (see Chapter 6). They encouraged the holding of two-day meetings wherever possible in order to accommodate expanding programmes which frequently included colloquia, discussions, special lectures and Symposia as well as free communications and demonstrations. They recommended closer discussion and collaboration with the Chemical Society in the areas covered by Biochemical Society meetings because the recent splitting of the *Journal of the Chemical Society* into sections resulted in two of these dealing with material which was appropriate to the *Biochemical Journal*. The recommendation of the Sub-Committee for more support for innovation in teaching and more interest in Biochemistry in industry, has been vigorously pursued so that today we have a thriving Professional and Educational Committee (Chapter 8). The Committee also agreed to transfer more work from the Honorary Secretaries to the Executive Secretary and his staff in Warwick Court.

From 1966 to the present day, the Main Committee has gradually changed its activity, becoming more in the way of a Council dealing with policy matters raised by advisory bodies. The first two were the Advisory Committee for Publications, formed in February 1963, which had evolved from a Sub-Committee set up to resolve differences between the Main Committee and the Editorial Board of the *Biochemical Journal*, and the Finance Sub-Committee, which was established in 1966. They were replaced by the Publications Board and the Finance Board, respectively, in 1973. A Meetings Board was also set up at this time. These Boards, together with the *Biochemical Journal* Editorial Board (which had been set up in 1944) and the Professional and Educational Committee (reconstituted in 1984 from the Professional and Educational Sub-Committee, itself reconstituted in 1977 from the original Professional Sub-Committee formed in March 1970) represent a very strong advisory input into the Main Committee. Cross-representation of the membership of the Boards ensured, in theory and frequently in practice, co-ordination of action. As Dr D. F. Elliott (Plate 4A), the Honorary Treasurer at the time, wrote: "This new organization fulfilled the essential need in the financial area on such crucial issues as the allocation of funds, the pricing for publications, the control of expenditure and the level of the membership subscription". All this will be elaborated on in Chapter 4. An unusual outcome of these developments was that in 1976 the work of the Meetings Board became superfluous and the members themselves recommended that their activities be suspended on the understanding that the situation would be considered in the following year. However, the Board still remains in abeyance. Recently a long-term planning group has been set up; its first report (November 1985) dealt with a number of urgent issues, none of them particularly long term. A completely new organizational plan for the Society is now (1986) before the Committee for consideration.

A summary of the major developments initiated by the Society from 1944 to 1985 is given in Table 3.6. Most of these

Table 3.6. Developments initiated by the Biochemical Society, 1944-1985

Year	Development	Year	Development
1944	Proposed formation of Biological Council	1969	BDH Awards in Analytical Biochemistry established
1947	Initiative for organizing First International Congress of Biochemistry	1970	Professional Sub-Committee established
1949	Post of Symposium Organizer established	1970	Heads of Departments conference formalized
1955	Post of Deputy Editor of *Biochemical Journal* established	1971	First Refresher Course held
1956	First Meeting with British Biophysical Society	1971	Boehringer Mannheim Travelling Fellowship introduced
1958	Hopkins Memorial Lecture	1972	Colchester Book Depot bought
1961	First Administrative Secretary appointed	1973	Finance Sub-Committee changed to Finance Board
1961	50th Anniversary of Founding of Society	1973	Publications Board established
1961	Jubilee Lecture established	1973	*Biochemical Society Transactions* introduced
1962	Joined with Medical Research Society in publishing *Clinical Science*	1973	Post of Honorary Careers Advisor established
1962	Jubilee Lecture established	1973	Meetings Board constituted
1963	Initiative taken for establishing a Federation of European Biochemical Societies (FEBS)	1976	Meetings Board disbanded
1963	Advisory Committee for Publications established	1976	Promotions Organizer post established
1963	Colworth Medal struck	1977	Professional Sub-Committee expanded to Professional and Educational Sub-Committee
1964	Post of International Secretary established (discontinued after formation of FEBS)	1978	Morton Lecture instituted
1964	Essays In Biochemistry first published	1978	Wellcome Trust Award for Research in Biochemistry Related to Medicine established
1964	First Group established	1981	*Bioscience Reports* published (transferred to a commercial publisher in 1985)
1964	David Keilin Memorial Lecture established	1981	Meetings charge introduced
1965	Ciba Medal struck and Prize established	1982	Post of Honorary Public Relations Official established to replace Promotions Organizer
1965	Unilever European Fellowship scheme initiated	1983	Junior Travelling Fellowships established
1966	Grants to student Biochemical Societies introduced	1983	Schoolteacher Fellowship established
1966	Incorporation under the Companies Act	1984	Professional and Educational Sub-Committee upgraded to Professional and Educational Committee
1966	Finance Sub-Committee established	1984	Krebs Memorial Scholarship founded
1966	Purchase of 7 Warwick Court	1985	Sponsorship of the Biochemical Society Exhibition in the Science Museum, South Kensington
1967	Separation of membership subscription from *Journal* subscription	1986	Meetings charge abandoned
1967	Harden Conference established		

are discussed in detail in appropriate sections of this and later chapters.

3.5 Biological Council

In March 1944 the Committee agreed to initiate discussions on the possible formation of a Biological Council, and recommended that the number of discussion meetings should be increased and that the teaching of Biochemistry should be assisted whenever possible.

With regard to the formation of the Biological Council, R. A. Peters (Fig. 3.11), F. G. Young (Fig. 3.12) and W. T. J. Morgan (Plate 4A) were asked to look into the matter and, after various soundings, a meeting of ten interested Societies was held at The Royal Society in September 1944. A memorandum was produced, mainly the work of Walter Morgan, for circulation and comment. At the meeting of the Society's Committee in September 1945 Morgan was able to announce the formation of the Biological Council. It was supported by contributions from constituent societies to the extent of £5 [£60] per annum or one guinea (£1.05) per 100 members "whichever sum be the lesser". The financial limit was soon raised to £10. The present contribution of the Society is £35.

Having helped to achieve the birth of the Biological Council the Society did not continue to take a great part in its activities. The Biological Council eventually spawned the Institute of Biology, a professional organization similar to the Institute of Chemistry (now subsumed within the Royal Society of Chemistry), a development which the Society's Committee viewed with limited enthusiasm. It did, however, eventually send a donation of £5 towards the foundation of the Institute.

Fig. 3.11. Sir Rudolph Peters, M.C., F.R.S. Hopkins Memorial Lecturer, 1958. Chairman of the Society Committee, 1928–1929, 1938–1939, 1952–1953. Honorary Member, 1967.

3.6 Anniversary Meetings

At the A.G.M. at UCL in March 1961, the fiftieth anniversary of the founding of the Biochemical Society was celebrated. A two-day symposium on 27, 28 March was held on "The Structure and Biosynthesis of Macro-Molecules". A conversazione mounted during the evening of 27 March in the North Cloisters of UCL was attended by 650 members. The event was supported by substantial donations from Industry. On the early evening of 28 March, a historic meeting was held in the Royal Institution when Sir Hans Krebs (Fig. 3.13) delivered the Third Hopkins Memorial Lecture on "The Physiological Role of Ketone Bodies". Apart from the intrinsic scientific merit of the lecture, Krebs described for the first time in public the details of his flight from Nazi persecution and paid a moving tribute to the friendship and help given him by Gowland Hopkins "when the Country of my birth proscribed me". The full text of the tribute is given in Krebs's autobiography [1]. On the Wednesday morning 34 Communications were presented in three separate sessions and the meeting ended with the A.G.M.

The Anniversary Dinner, held in the New Refectory of UCL, was well attended and the Society was honoured by the presence of many scientific guests including the President of The Royal Society, Sir Howard Florey, representatives of 20

Fig. 3.12. Sir Frank Young, F.R.S. Honorary Secretary, 1940–1943. Chairman of the Editorial Board, 1942–1946. Chairman of the Society Committee, 1953–1955. Honorary Member, 1979.

Fig. 3.13. Sir Hans Krebs, F.R.S. Nobel Laureate, 1953. Hopkins Memorial Lecturer, 1961. Honorary Member, 1967.

Fig. 3.14. Professor J. N. Davidson, C.B.E., F.R.S. Honorary Secretary, 1947–1951. Chairman of the Society Committee, 1961–1963.

Fig. 3.15. Professor R. A. Morton, F.R.S. (Chairman of the Society, 1959–1961. Honorary Member, 1966) being presented with a special leather bound copy of his *The Biochemical Society: its History and Activities, 1911–1969*, by Professor G. A. D. Haslewood, Chairman of the Society, 1969–1971.

biochemical societies, the three then extant Honorary Members, Sir Rudolph Peters (Fig. 3.11), Sir Henry Dale (Fig. 2.13) and Sir Charles Harington (Fig. 2.7), and two original members, Sir Charles Lovatt Evans (Fig. 2.13) and G. W. Ellis. The principal guest was Lord Hailsham, then Secretary of State for Science, who replied to the toast of "the Guests" proposed by Professor J. N. Davidson (Fig. 3.14). Professor Marcel Florkin (Belgium) and Academician Oparin (U.S.S.R.) also replied. Sir Howard Florey proposed the toast of "the Society" and Professor R. A. Morton replied.

The 75th Anniversary, which is the *raison d'être* of this book, was celebrated by special events spread over the whole of 1986. Details are given in Chapter 8.

The 500th meeting of the Society was marked by the publication of *The Biochemical Society: its History and Activities 1911–1969*, prepared by R. A. Morton. At the dinner associated with the meeting, Professor Morton was presented with a specially bound copy of his *History* (Fig. 3.15). The then President of the U.S. National Academy of Sciences, Dr Philip Handler, a distinguished biochemist, and his wife were special guests of honour as was Mrs Shirley Williams, then Minister at the Department of Education and Science. Professor G. A. D. Haslewood (Fig. 3.15), Chairman of the Society at the time, recalls that it is not only Britishers who succumb to the famous Williams' charm. Handler, at one point during the evening, turned to Haslewood, "Tell me" he said, "do you have many politicians like that?" [2].

A special two-day symposium was arranged for this anniversary meeting, on "British Biochemistry Past and

Table 3.7. Speakers and the titles of their lectures for Biochemical Society Symposia no. 30, *British Biochemistry Past and Present*, **held on the occasion of the 500th meeting of the Society**

Molecular biology

Chairman's Introduction
  J. N. DAVIDSON
Some remarks on the history of molecular biology
  J. C. KENDREW
The development of crystallographic enzymology
  D. C. PHILLIPS
The primary structure of proteins
  B. S. HARTLEY
Retrospect on the biochemistry of plant viruses
  N. W. PIRIE

Immunology

Chairman's Introduction
  J. R. MARRACK
The nature of the immune response
  J. H. HUMPHREY
The structure and combining specificity of antibodies
  R. R. PORTER
Carbohydrate structure responsible for antigenic specificity
  W. T. J. MORGAN

Intermediary metabolism

Chairman's Introduction
  A. NEUBERGER
Intermediary metabolism of animal tissue between 1911 and 1969
  H. A. KREBS
Lessons learnt from small molecules
  G. POPJAK
The role and maintenance of the tricarboxylic cycle in *Escherichia coli*
  H. L. KORNBERG

Separation methods

(Chairman: A. J. P. Martin)

A retrospect on liquid chromotography
  R. L. M. SYNGE
Methods for determining sequences in RNA
  F. SANGER and G. G. BROWNLEE
The development of gas–liquid chromatography
  A. T. JAMES

Present". It was a survey, inevitably limited, of achievements between 1911 and 1969. It was published as No. 30 in the Biochemical Society Symposia series. As the Editor wrote in the preface: "Scientists are pre-eminently international in outlook and some of our contributors were a little apprehensive of the emphasis which was placed on British Biochemistry. However, this was a unique occasion and we can be well satisfied with the contributions made to the subject from the U.K. during the past 58 years". Consideration of the list of the speakers and the titles of their lectures (Table 3.7) justifies this comment.

3.7 General Pattern of Ordinary Meetings

Despite considerable difficulties the Society's Officers maintained a viable programme of meetings during the Second World War. Instead of the agreed eight meetings per year, the numbers during 1940–1945 varied between five and six with about 50 Communications and 10–15 demonstrations per year. Only once was a meeting cancelled because of lack of a sufficient number of communications. This limited programme was kept alive only by the heroic efforts of the Officers. Professor Walter Morgan (Plate 4A) recalls that "things were pretty grim ... and the programmes show that at some of the meetings very few [papers] were presented and I remember the difficulty of getting the printing done and the uncomfortable journeys to Scotland to see that 4–5 papers (only) were given and read the 'Minutes of the last meeting'. All for the record! But we had enjoyable small dinners and kept in touch with other members" [3].

The number of meetings per year was restored to eight after the War and the size of the meetings rose slowly but steadily until in 1956–1957 the number of Communications was 196 and the number of demonstrations 33. At this time the attendance at meetings varied between 100 and 300, most attracting over 200 participants.

Discussion meetings which had begun again in 1941 (see the preceding chapter) continued to flourish to such an extent that considerable support emerged for the publication of the proceedings at such meetings. In September 1946, the Editorial Board of the *Biochemical Journal* indicated that they did not wish to incorporate these proceedings into the *Journal*. After consideration of the estimated cost of publishing the discussions separately it was decided to launch a new publication of discussion meetings under the title of *Biochemical Society Symposia* with R. T. Williams (Fig. 3.16) as Symposium Organizer. He successfully set the *Symposia* on the right path and served the Society loyally for 10 years in the capacity of Symposium Organizer.

After a slow start the *Symposia* became popular and sold well; some, for example, "Partition Chromatography and its Application to Biochemical Problems" (1948), were particularly successful. It was agreed that as the *Symposia* were so well established the major contributors would receive a free copy of the proceedings but no reprints and that the discussions following the main papers would not be recorded. The first two rules still apply but certainly by volume 12 (1954) selected contributions to the discussion of main papers were being published. When Professor Williams retired it was agreed that the Symposium Organizer should be given greater status and become an *ex officio* member of the General Committee; it was also agreed that the tenure of office should be limited to 7

Fig. 3.16. Professor R. T. Williams, F.R.S. Symposium Organizer, 1945–1955.

Table 3.8. Symposium Organizers

	Dates
R. T. Williams, F.R.S.	1945–1955
E. M. Crook	1955–1958
J. K. Grant	1958–1964
T. W. Goodwin, C.B.E., F.R.S.	1964–1970
R. M. S. Smellie	1970–1976
P. B. Garland	1976–1980
C. E. Phelps	1981–1984
J. Kay	1985–

years. Two symposia per annum were to be aimed for, but this was not always feasible. Under a number of Symposium Organizers (Table 3.8), the *Biochemical Society Symposia* have become an established and scientifically successful activity of the Society.

To return to Discussions for a moment, it is interesting that "The Chemical Basis of Cell Structure" (1945) involved the participation of a number of French biochemists and the meeting was marked by the presentation of a Pasteur Medal to the Society by Professor R. Fabre on behalf of the French Biochemical Society. This was probably the first occasion after the 1939–1945 War that formal contact between British and European biochemists was made. What eventually developed from this contact, and the Society's part in the formation of IUB and FEBS, is given in full in Chapter 7. The topic of the first post-War joint Discussion with another Society (Society of General Microbiology), was "Quantitative Biochemical Analysis by Micro-biological Response" (1946).

When the Symposia became established it was agreed that they could be held outside Oxford, Cambridge and London. The first such meeting was on "The Biochemistry of Fish" at Liverpool in 1949. It is also interesting to note that around this time the Symposium Organizer could not find sufficient speakers for a proposed symposium on "Plant Biochemistry".

In 1967 the first Symposium to be arranged by the Society specifically to pay tribute to one biochemist was held in Oxford. It will come as no surprise that the biochemist was Sir Hans Krebs (Fig. 3.13) and that the title of the Symposium was "Metabolic Roles of Citrate". As we shall see later, *Biochemical Society Transactions* has taken over the role of publishing the proceedings of meetings organized to salute the achievements of members of the Society. The organizing of a Symposium sometimes resulted in an unexpected and fruitful scientific spin-off. The Symposium held on Neurochemistry in 1951 was, according to Professor H. McIlwain [4], of great importance in catalysing activities which eventually led to the formation of the International Society of Neurochemistry (ISN).

Although Symposia helped to focus meetings of the Society on some topical research development, the fact that they were only annual events precluded them from making a significant contribution to a problem which was rapidly developing in the 1960s. The problem was to reconcile the demand for more time to present original work with the rapid increase in specialization within Biochemistry. Many specific papers presented at a meeting were of direct interest only to a small number of the biochemists participating in the meeting. Often the audience for some presentations was the Chairman, the speaker's supervisor and the next speaker and his supervisor. The first attempt to deal with this problem was the decision to organize Colloquia to be held at every meeting, the topics to be of particular interest to the Host Department. The first of these Colloquia was held in Liverpool in January 1964 and the topic chosen was "Aspects of Vitamin A Function". However, even this development, excellent as it was, was not sufficient to deal with the explosive expansion of Biochemistry into all areas of Biology, particularly into Molecular Biology. The Society's response to this was the introduction of Groups. It is not an exaggeration to say that this development, more than any other single activity, saved the Society from the possibility of extinction by the uncontrolled splintering off of new societies. This important development is dealt with in detail in Chapter 5. There are now some Group Colloquia and Poster Sessions at every meeting of the Society. In addition a Society/Host Colloquium is now also arranged for each meeting so that the Host Department is free to choose its own subject for discussion.

As oral Communications became less and less attractive to members, an innovation which the Society took up with enthusiasm early on and made an outstanding success, was the Poster Session. Such sessions are now an important part of all meetings; indeed they appear to have insinuated themselves into meetings of almost all scientific societies. Presentation of new material as a Poster provides an ideal environment for informal but informed discussion of the work on a one to one basis and leads not only to greater understanding of the problems investigated, but to many contacts and frequently important collaboration mutually advantageous to presenter and discussant. The first Society Poster Session, although not designated as such, was held at the MRC laboratories at Carshalton in 1967 on the initiative of Dr W. N. Aldridge (see Chapter 6). At this meeting authors 'demonstrated' their results in an informal way with the aid of prepared cards. Such sessions soon became part of every meeting and Fig. 3.17 recalls one of the early meetings held at UCL in 1970. It is difficult to decide just when the term 'Poster' came into general use. Oral presentation of free Communications was formally discontinued following a Committee decision of 25 November

Fig. 3.17. One of the early Poster Sessions held at UCL in 1970.

1982; furthermore Groups were encouraged to extend their present practice of incorporating Poster material into specific oral sessions, e.g. round table discussions, organized as part of their Colloquia.

Oral presentations in separate pre-doctoral sessions have, however, been a successful aspect of the Irish Group activities (see Chapter 5). An attempt in 1986 to establish such a meeting in the U.K. on the grounds of giving pre-doctoral students experience in presenting their work to a critical audience, failed through lack of support.

During the past 25 years or so, many special lectures have been endowed (see section 3.10) so that there are very few meetings which do not include a named lecture in addition to all the other attractions just described, as well as some type of commercial exhibition. The outline programme (Table 3.9) of a meeting which took place in Cardiff in 1985 shows how wide-ranging and attractive the modern meetings are and how rapidly they have progressed since the later 'forties when they consisted of about 10–15 Communications and two or three demonstrations, all taking place in one afternoon (usually Saturday) session, with a break for tea. It is also obvious that Free Communications on General Topics are today very much a minor part of most meetings. Because of the size and complexity of current meetings the number per year has been reduced from eight to four or five. This concentration of effort makes the meeting much more economical of time and money and much more worthwhile scientifically.

The Oxford meeting of the Society in July 1985 broke all records. Attendance was greater than at some FEBS meetings and many intending visitors could not be found accommodation.

Table 3.9. A summary of the schedule for the 613th Biochemical Society Meeting, University College Cardiff, 20–22 March 1985

	John Pryde Lecture Theatre	Physiology A Lecture Theatre	Shared Lecture Theatre	Deck Laboratory/Physiology Foyer/Shared Lecture Theatre Foyer
WED. MARCH 20	Carbohydrate recognition systems in animals (carbohydrate Group Colloquium) 09.30–17.15 COLWORTH MEDAL LECTURE 17.30	Structure and activity of aspartic proteinases (Molecular Enzymology Colloquium) 09.30–17.15	Hormones and receptor-mediated internalization (Hormone Group Colloquium) 09.15–12.40	Posters. Free Commun. A. General topics I 13.15–14.00 B. Molecular Enzymology Group 13.15–14.00 C. Carbohydrate Group 13.50–14.50 D. Hormone Group 14.00–15.30
THURS. MARCH 21	Cellular proteolysis (Society/Host Colloquium) 09.00–17.15 MORTON LECTURE 17.30	The Pharmacological Biochemistry of stimulus–response coupling (Pharmacological Biochemistry Group Colloquium) 09.00–17.15	Recent applications of HPLC to Biochemistry (Techniques Group Colloquium) a.m. Recent advances in HPLC column technology 13.45–17.00	Posters. Free Commun. A. Society/Host Colloquium 13.00–14.00 B. Pharmacological Biochem. Group 13.15–14.00 C. General Topics II. 13.15–14.00
FRIDAY MARCH 22	Lung surfactants (Lipid Group/Membrane Group Joint Colloquium) 09.30–16.15	Recent application of HPLC to Biochemistry (Techniques Group Colloquium) A. Recent advances in HPLC analysis of small molecules 09.00–12.30 B. Recent advances in the HPLC analysis of middle molecules and large molecules 14.00–16.30	Newer aspects of the Biochemistry of receptors (Neurochemical Group Colloquium) 09.00–17.00 **Anatomy Lecture Theatre** Wet or dry: the fate of the biochemical practical (Education Group) Informal Session a.m.	Posters. Free Commun. A. Education Group 11.00–11.30 B. Neurochemical Group 13.15–14.00 C. Techniques Group 13.15–14.00 D. Lipid/Membrane Group 13.30–14.00 E. General Topics 13.15–14.00

Even with all these successful activities in train the Committee were still worrying about some members' interests not being effectively covered. So in 1984, on the advice of Professor R. H. Burdon, then Honorary Meetings Secretary, they announced the inauguration of Special Colloquia thus:

"… it has been pointed out from time to time that some Members' interests are not adequately covered by any of the Groups. Alternatively, it has sometimes been the case that growth areas are not directly in the field of interest of any single Group. This, of course, can often be remedied by Joint Group Colloquia. However, such developments can often be ignored because they are neither one thing nor another. More simply, they sometimes do not receive support because Groups with limited financial resources may promote more 'popular' aspects of their subject area. To compensate a little for this, and to promote a more flexible approach, it is proposed to mount one, or two, Special Colloquia annually. Special Colloquia would be one, or halfday, events, held as part of normal Society Meetings. However, they would be organized on a 'one-off' basis by *ad hoc* groups of Society Members with help from the Meetings Office. Specifically, the subject areas must be in areas not adequately covered by the present Group structure. Indeed, where there is any doubt, the relevant Group Committees would have to be consulted and give their approval.

"Any group of Society Members (5 to 10), therefore, who feel they would like to organize a Special Colloquium with the above constraints in mind, are invited to present their case in writing to the Honorary Meetings Secretary. Depending on suitability of their case, financial backing from the Society and timing of a Special Colloquium will be discussed in relation to other Society activities."

The great surge of activity in the early 'sixties was achieved only by the extraordinary devotion and hard work of the Honorary Meetings Secretaries; their activities frequently bordered on the phrenetic. Thus W. J. Whelan (Fig. 3.10): "… on the press day I set aside practically the whole day to edit the abstracts and prepare the associated announcements. By continuous work in this way I could cut the overall time to a minimum but I usually found myself working against the clock and my wife became more used to a regular midnight drive around London, first to put the abstracts on the overnight train to Cambridge from Liverpool Street and then over to Paddington to catch the corresponding train to Castle Cary for the announcements. On one occasion, when it seemed that everything would be held up in the Christmas postal delays, I drove to Cambridge and transported the abstracts by car to Castle Cary".

H. R. V. Arnstein (Plate 3A), Whelan's successor as Honorary Meetings Secretary, found that the International meetings were particularly frenzied: "We were moving house in

1963 on the day of the deadline for sending everything to the printers. I had received over 100 abstracts, many of them within a day or two before they had to be sent to the press, since papers often arrived late, particularly from abroad. Fortunately, Bill Whelan had kindly agreed to help with the editing and we managed to get the paperwork for the meeting done on a couple of packing cases in my study while our furniture was being unloaded all around us". There was perhaps one advantage which we do not enjoy today: "... in those days the last mail, even as far away from the centre of London as Mill Hill where I lived, was at 11 p.m. and one could rely on delivery the next day".

It was around this time that the burden for Honorary Meeting Secretaries became intolerable and the Society appointed a full time Meetings Officer. Miss Doris Herriott (Plate 2A) was appointed in 1961 and since then her commitment to the Society has been as great as that of the Honorary Officers, whom she has served with such enthusiasm. But with the developments just described the workload exceeded even her capacity and, as indicated in section 3.2, an Assistant Meetings Officer was appointed in 1985.

The Society's response to the requirement of the rapidly burgeoning subject by expanding its scientific activities in the form of larger and more intricate meetings, as exemplified in Table 3.5, has been generally of inestimable benefit. This has been achieved at the cost, which is particularly apparent to older members, of the Society having become a much more impersonal organization. This is an inevitable effect of increase in size. Gone are the days when the Society was indeed a club where all members knew each other and on the whole could understand each others' papers. Gone are the days of 'characters' whose presence at a meeting was always obvious and who generally 'performed' as expected. One such flamboyant character was A. L. Bacharach who, beside being an industrial biochemist, was a reviewer of detective novels (he claimed to have read one a day) and a true musicologist (he edited *The Musicians' Companion*, which has recently appeared in a new updated edition). He was also a very elegant dresser; Professor G. A. D. Haslewood (Fig. 3.15) recalls that "Earl King (Fig. 3.2) and others held sweeps on what colour of tie Bacharach would wear at the next meeting. A particularly acrimonious debate took place when the Committee proposed that the Proceedings ... should be edited (see Chapter 5). 'Not even the combined sagacity of the Editorial Board and its referees could ensure that some revolutionary discovery would not be left unannounced if the Proceedings were edited' said Bacharach in a speech that ensured the motion's defeat". Another typical comment recalled by Haslewood was on a paper read during the Second World War, at Hampstead: "What I admire about this paper" said Bacharach, "is the author's courage, for these findings had already been published in … 1916!".

3.8 Proceedings, Agenda Papers, *Bulletin*

As we shall see from Chapter 6 where the history of the *Biochemical Journal* is considered, it was becoming clear in the early 'sixties that the membership subscription could not carry free provision of the *Biochemical Journal*. This meant that the only tangible result of membership would be the receipt of the Agenda Papers, which, as Abstracts of Proceedings, had been precirculated since 1938. The Committee felt that better value for money should be apparent and decided that the style of the Agenda Papers should be changed to match that of the *Biochemical Journal* and that the Proceedings, previously published in the *Biochemical Journal*, would become part of the enlarged Agenda Papers. This arrangement worked satisfactorily until the need for a house journal to keep members informed of the many activities of the Society and its members, resulted in the metamorphosis of the Agenda Papers into the *Biochemical Society Bulletin*. The first number appeared in February 1979 and contained an article by Dr Peter Mitchell, Nobel Laureate. The *Bulletin* has now become accepted as a well established aspect of the Society's activities, although Colloquia at meetings are now recorded by title only. Since 1985, thanks to the efforts of the Honorary Meetings Secretary, Communications have been separated from the *Bulletin* and are printed separately in booklet form from authors' camera-ready copy, and mailed to members with the *Bulletin*. This allowed much more flexibility in arranging the programmes of meetings which, as indicated above, were becoming very large and increasingly complex. The booklet for the Oxford meeting in July 1985 consisted of 153 pages with generally three abstracts on each page.

In spite of the publicity which the Society willingly generates through its meetings, it still insists that the proceedings of the meetings are private and must be reported to the Press only after clearance with the Executive Secretary and the Honorary Public Relations Officer. This sensible rule is mainly to protect the participants from misquotation in the non-scientific press. It stems from 1926 when Sir Charles Harington was incensed to find a garbled version of his Communication in a national newspaper.

This section cannot end before acknowledgment of the debt the Society owes to Heads of University Biochemistry Departments who, together with their own colleagues, shoulder the considerable burden of setting up meetings in their Departments. Even with the expert help of the Meetings Office a great deal of work is always necessary at the grass roots level. Indeed the popularity of meetings meant that sympathetic industrialists had to withdraw from hosting meetings because of lack of sufficient lecture room space. The success of some meetings can occasionally overwhelm Departments. One elder statesman once called an Honorary Meetings Secretary over the coals after an International meeting had disrupted his entire department for more than a week in July. The Secretary was told to arrange the next International meeting in September when a reduced number of participants could be expected. This was accordingly done but the attendance was so poor that the complaint then was that there were too few participants!

3.9 Travel Funds

In the context of the financial support of present day activities in numerous walks of life it will come as a surprise to many members that the business of running the Society was, until 1942, carried out with out-of-pocket expenses rarely being available. In 1942 the Committee eventually grasped the nettle and agreed that Committee members be paid third class travelling expenses if their journey to the meeting place exceeded 30 miles. Nowadays the members of the various Committees are paid first class rail or economy airfare and the Government recommended subsistence allowance.

Since the Second World War, the need for travel funds to allow scientists to present papers at Conferences both at home and overseas has eventually been accepted by the Research Councils and the Universities. The Society was one of the first scientific societies in the field in providing their own travel funds and although originally the funds available were quite modest, the grants now provided represent a considerable proportion of the overall travel monies available to members.

3.9.1 IUB Congress Funds

The first travel funds for members were raised by Professor F. L. Warren (Honorary Secretary, 1953–1958), who persuaded the Wellcome Trust to support the attendance of 25 Members at the Fourth International Congress of Biochemistry in Vienna in 1958. This support by the Wellcome Trust to the extent of £500 p.a. lasted until the Amsterdam Meeting in 1965. Government funds for travel became available to The Royal Society around 1958 and allocations to attend international Congresses organized by members of ICSU are now made by the Council of The Royal Society after consideration of submissions from the various National Committees. In the case of Biochemistry the funds available from The Royal Society are pooled with those provided by the Society and allotted to appropriate participants by a joint *ad hoc* Committee of the Society and The Royal Society. It is an open secret that The Royal Society has always been impressed by the amount of money collected for travel by the Society. On the other hand, the Society can also be pleased that the submissions of The British National Committee for Biochemistry have always been looked on extremely favourably by The Royal Society.

3.9.2 FEBS Congress Funds

As Congresses organized by FEBS became more important on the biochemical scene the Society decided to support applications from members to attend such Congresses. FEBS funds are also provided to support these meetings.

3.9.3 General Travel Funds

Following the financial surplus made in 1982 (see Chapter 4) the Committee of the Society increased the capital of the Travel Fund so as to allow support for attendance at meetings with a substantial biochemical content other than the major FEBS meeting and IUB Congresses. At the moment no single grant is likely to exceed £300. Eligibility for such grants has been made as wide as possible; they are available to members resident in any part of the world to travel to meetings anywhere in the world. There is only one exception: U.K. residents travelling to meetings in the U.K. are not eligible for a grant.

3.9.4 Student Travel Funds

Yet another source of support became available when the Committee agreed to provide Heads of University Departments of Biochemistry and Related Sciences with a grant equivalent to £20 (raised to £30 in 1985) per annum for each full-time student member of the Society registered in the Department. The money can be allocated amongst the students as the Head of Department thinks fit provided that it is all used to cover travelling expenses incurred by the students in attending meetings of the Society. This fund is in addition to that which has been available annually for some years to Student Biochemical Societies for the expenses of a named "Biochemical Society Lecturer". This has been well appreciated and has given Student Societies great encouragement.

Up to date details of available travel funds can always be found in the latest Society Yearbook.

3.10 Medals and Named Lecturers

The portfolio of medals and named Lectures which the Society now holds is impressive. They divide naturally into two groups: firstly those intended to honour famous British bio-

Fig. 3.18. The Hopkins Memorial Medal.

chemists who have contributed outstandingly to the development of the subject; secondly, those sponsored by Industry to recognize outstanding work, frequently with emphasis on the achievements of younger workers.

3.10.1 Hopkins Memorial Lecture

In 1956 there appeared before the Committee a request to establish a memorial lecture to be presented biennially by an outstanding British or French biochemist, a medal to be provided by the sponsors. After much discussion the Committee decided "that since the institution of such a medal and lectureship would involve a completely new departure on the part of the Society it would not agree to the request". On the face of it this would appear to have been a very conservative reaction of the Committee. In fact it was not so; the proposal was not really appropriate for the Society to pursue. However, the ideas of memorial lectures and medals were now in the air and a proposal at the 1957 A.G.M. by Dr T. S. Work (Plate 1A) that a Hopkins Memorial Lecture be established was agreed unanimously. In this case the Committee reacted very positively and formally instituted the lectureship on 14 February 1958. They allocated £2000 [£25,000] for the purpose, had a medal cast by Pinches and drew up the rules for the award with such speed that it was possible to make the first award in 1958. Very appropriately the first recipient was Sir Rudolph Peters (Fig. 3.11), who received his medal and presented his lecture in April 1959. The likeness of Hopkins which appeared on the medal (Fig. 3.18) is from a pencil sketch prepared by Pinches from three photographs obtained by Dr F. A. Robinson (then Treasurer, see Chapter 4) from Hopkins's daughter Mrs E. Holmes; the final sketch was approved by his family. The recipients of the medal, many of whom had worked with or been associated with Hopkins, are listed in Table 3.10.

3.10.2 The Jubilee Lecture

The Committee, in organizing events to commemorate the Society's 50th Anniversary, agreed to establish a Jubilee

Table 3.10. Recipients of the Hopkins Medal

1958	Sir Rudolph Peters, F.R.S.	1971	F. Sanger, O.M.C.H., F.R.S.
1960	A. Neuberger, C.B.E., F.R.S.	1973	M. F. Perutz, C.H., C.B.E., F.R.S.
1961	Sir Hans Krebs, F.R.S.	1975	E. Racker
1963	L. F. Leloir, Foreign Mem. R.S.	1977	R. R. Porter, C.H., F.R.S.
1965	A. Szent-Gyorgi	1979	J. Porath
1967	H. A. Barker	1981	F. Gibson, F.R.S.
1969	F. J. W. Roughton, F.R.S.	1983	E. G. Krebs
		1986	C. Milstein, F.R.S.

Lecture to be given biennially, alternating with the Hopkins Lecture. The lecturer, who receives an honorarium, is required to lecture on his chosen field both in London and at a suitable centre outside London. The first lecture was delivered in 1962 by P. C. Zamecnik. Since 1978 the lecturers have also received the Harden Medal. The Jubilee lecturers who have up to now been elected are listed in Table 3.11.

Table 3.11. Jubilee Lecturers

1962	P. C. Zamecnik	1974	J. H. Quastel, C.C., F.R.S.
1964	E. Lederer	1976	A. Kornberg, Foreign Mem. R.S.
1966	F. Lynen, Foreign Mem. R.S.	1978	C. de Duve, Honorary Member
1968	H. G. Khorana, Foreign Mem. R.S.	1980	S. J. Singer
1970	A. L. Lehninger	1982	H.-G. Hers
1972	C. B. Anfinsen	1985	A. Klug, F.R.S.
		1987	M. Z. Atassi

3.10.3 The Keilin Memorial Lecture

The Society was in 1963 asked to administer a fund raised by friends and colleagues of David Keilin, F.R.S., whose fundamental work on cytochromes is now classical Biochemistry. The Committee readily agreed to this and a biennial Keilin Memorial Lecture was instituted in January 1964. The lecturer also receives a medal. Those who have delivered this lecture are listed in Table 3.12.

Table 3.12. Keilin Memorial Lecturers

1964	A. Lwoff, Foreign Mem. R.S.	1974	E. C. Slater, F.R.S.
1966	B. Chance, Foreign Mem. R.S.	1976	S. M. E. Magnusson
1969	M. Eigen, Foreign Mem. R.S.	1980	J. Kraut
1970	E. Margoliash	1983	M. G. Rossmann
1972	R. J. P. Williams, F.R.S.	1985	H. Beinert
		1987	R. Huber

3.10.4 The Morton Lecture

The latest named lecture was instituted in 1978 to commemorate the achievements in fat-soluble vitamin Biochemistry of the late Professor R. A. Morton, F.R.S. (Fig. 3.15), Johnston Professor of Biochemistry at the University of Liverpool 1944–1966. The Biochemical Society agreed to administer the funds, which were the result of an appeal by Morton's colleagues and friends. The lecture is given biennially, once in the University of Liverpool and once at an ordinary meeting of the Society. The lecturer should have made outstanding contributions to lipid biochemistry. Four lectures have so far been presented, by L. L. M. Van Deenen,

Fig. 3.19. The Colworth Medal.

H. F. DeLuca, T. W. Goodwin and H. Rilling, in 1979, 1981, 1983 and 1985, respectively. The 1987 lecturer will be J. N. Hawthorne.

3.10.5   CIBA Medal and Prize and Colworth Medal

These two medals are considered together not because they do not deserve to be treated singly, in fact quite the contrary, but because they were the Society's first prizes to be endowed by Industry. The Medals came into being by an interesting concatenation of events. The Honorary Secretary in 1962, Professor H. R. V. Arnstein (Plate 3A) had worked at the National Institute for Medical Research together with Dr A. T. James and Dr D. F. Elliott (later Honorary Treasurer of the Society, see Chapter 4). The two last named moved to Industry and Arnstein soon began to persuade them to interest their firms in the Society. The outcome of the efforts of Dr James and the late Dr H. Wilkinson, then Director of the Unilever Research Laboratory at Colworth House, was the financing in 1963 of the Colworth Medal (Fig. 3.19).

This is awarded annually to a British biochemist, not normally over the age of 35. The recipient is expected to give a lecture to a meeting of the Society and to repeat it at the Unilever Research Laboratories at Colworth House. Over the years this award has gained in prestige and is now generally accepted as the highest accolade which can be bestowed by the Society on a young British biochemist. A glance at the list of recipients (Table 3.13) impressively bears this out and, incidentally, emphasizes the acumen of the awarding Committees.

The CIBA Medal (Fig. 3.20) and prize was inaugurated in 1964 by the CIBA Research Laboratories, Sussex, as a result of the good offices of Dr Elliott, who was their Research Director. It is awarded each year in recognition of outstanding research in any branch of Biochemistry. The award is for work carried out in the U.K. but candidates can be of any nationality. A prize is associated with the Medal and the recipient is

Fig. 3.20. The CIBA Medal.

Table 3.13. Colworth Medallists

1963	Sir Hans Kornberg, F.R.S.	1975	W. J. Brammar
1964	J. R. Tata, F.R.S.	1976	G. G. Brownlee, F.R.S.
1965	J. B. Chappell	1977	P. Cohen, F.R.S.
1966	Sir Mark Richmond, F.R.S.	1978	T. E. Hardingham
1967	L. J. Morris	1979	R. A. Laskey
1968	P. B. Garland	1980	R. A. Flavell, F.R.S.
1969	G. K. Radda, F.R.S.	1981	T. H. Rabbitts, F.R.S.
1970	D. A. Rees, F.R.S.	1982	D. M. J. Lilley
1971	A. R. Williamson	1983	E. Oldfield
1972	J. M. Ashworth	1984	M. D. Houslay
1973	J. C. Metcalfe	1985	A. J. Jeffreys, F.R.S.
1974	D. R. Trentham, F.R.S.	1986	G. P. Winter

Table 3.14. Ciba Medallists

1965	Sir John W. Cornforth, C.B.E., F.R.S. ⎫ joint	1975	E. F. Hartree
	G. J. Popjak, F.R.S. ⎭ award	1976	S. V. Perry, F.R.S.
1966	R. R. Porter, C.H., F.R.S.	1977	C. Milstein, F.R.S.
1967	D. M. Blow, F.R.S.	1978	J. R. Quayle, F.R.S.
1968	W. J. Whelan	1979	J. B. Gurdon, F.R.S.
1969	T. W. Goodwin, C.B.E., F.R.S.	1980	S. Brenner, F.R.S.
1970	Sir David Phillips, F.R.S.	1981	I. H. M. Muir, C.B.E., F.R.S.
1971	D. H. Northcote, F.R.S.	1983	G. K. Radda, F.R.S.
1972	R. T. Williams, F.R.S.	1984	Sir Philip Randle, F.R.S.
1973	P. D. Mitchell, F.R.S.	1985	E. A. Barnard, F.R.S.
1974	E. Kodicek, C.B.E., F.R.S.		

expected to deliver a lecture. The recipients of this award, more senior than those awarded the Colworth Medal, are listed in Table 3.14 which again makes impressive reading.

3.10.6 BDH Award in Analytical Biochemistry

Towards the end of his term as Honorary Secretary Professor Arnstein was approached by Dr Bayley of BDH Ltd. He asked whether the Committee would be interested in accepting a donation for an award to be given in recognition of work leading to the development and application of a new reagent or method. The Committee accepted this generous offer and a triennial award available to Members of the Biochemical Society was instituted in 1969. The first winner, Professor B. H. Hartley, appropriately gave his lecture at the 50th meeting of the Society. A full list of awardees is given in Table 3.15.

Table 3.15. Recipients of BDH Award in Analytical Chemistry

1969	B. S. Hartley, F.R.S.
1972	J. E. Scott
1975	J. Landon
1978	H. R. Morris
1981	E. M. Southern, F.R.S.
1984	J. Chayen
1986	D. Robinson

3.10.7 Wellcome Trust Award for Research in Biochemistry related to Medicine

In 1977 the Wellcome Trust generously offered a biennial award of £500 for distinguished research leading to new advances in medical science. The award is intended to recognize the achievements of biochemists who are under the age of 45 at the time of the award. The research attracting the accolade has to have been carried out in the U.K. or Ireland during the seven years preceding the date of the nomination.

The recipients up to date have been D. J. H. Brock (1978), K. B. M. Reid (1981), R. Williamson (1984) and G. G. Brownlee (1986).

The awardees are expected to lecture at a meeting of the Society and to prepare a manuscript for publication in *Biochemical Society Transactions*.

3.11 Fellowships and Scholarships

It is only right and proper that the Society should honour outstanding achievements by awarding medals and lectureships but it is equally if not more important that it should actively support young promising research investigators. In achieving this aim with a series of Fellowships it has been generously supported by Industry.

3.11.1 Unilever (Short-Term) European Fellowships

Help in this direction was again forthcoming from Unilever, who in 1965 established two Fellowships of £1500 each to be awarded annually, one for a British biochemist to work in a laboratory in Continental Europe and one for a European national to work in a U.K. laboratory. The only condition attached to the award of the Fellowship was that any publication arising from work carried out during the period of the award should carry the statement that the author was a 'Unilever Fellow of the Biochemical Society'. Over the years this scheme has been extremely successful, particularly in providing the possibility of European biochemists to carry out research in the U.K.; recently, however, demand for these Fellowships has been falling off and reconsideration of the situation has resulted in Unilever suggesting that the original Fellowships be replaced by a number of short-term Fellowships of up to £1000 (air fare plus £25 per day subsistence) each to fund short-term research visits. The following activities fall within the scope of the amended scheme: (i) training in new techniques; (ii) use of special research facilities; (iii) initiation of scientific collaboration and (iv) support of such collaboration in the absence of other funding. Holders of such Fellowships are expected to submit a report suitable for publication in the *Bulletin*, within two months of completing their Fellowships.

3.11.2 Boehringer–Mannheim Travelling Fellowship

The Boehringer Corporation (London) p.l.c. have provided, in honour of Sir Hans Krebs, £500 per annum since 1967 for a number of travelling Fellowships to allow younger biochemists (not over the age of 30) to spend short periods in another laboratory or to attend summer schools which would allow them to obtain training or experience not available in the U.K.

3.11.3 Biochemical Society Junior Travelling Fellowships

In 1983 the Committee decided to match the Boehringer funds so that it was possible to support even more young biochemists. Awards from this fund were designated 'Biochemical Society Junior Travelling Fellowships'.

3.11.4 Krebs Memorial Scholarship

An appeal was launched in 1982 for funds to be used to celebrate the life and work of Sir Hans Krebs (Fig. 3.13) by instituting a post-graduate Scholarship in Biochemistry or an allied biomedical science in any British University. An extremely generous response by the biochemical fraternity quickly made the Scholarship a reality and the first award was made in 1984. The Scholarship is unique in that it is primarily intended to help those whose careers have been interrupted for non-academic reasons beyond their control and/or who are unlikely to qualify for an award from public funds. This clearly reflected Krebs's continued interest in such unfortunate persons of whom he himself was one in the 1930s. The Scholarship covers a personal maintenance grant at an appropriate level, all necessary fees and a Research Training Support Grant. The last-named is the sum paid by the Research Councils to a Department which has been awarded one of the Research Training Studentships. The Krebs Scholarship is awarded for one year in the first instance but can be renewed up to a maximum period of three years. It is awarded in alternate years. The first award, for the academic session 1984–1985, was made to Mrs Marvash Tavassoli, a second year Ph.D. student at the University of Sussex.

3.11.5 Schoolteacher Fellowships

Following a recent recommendation from the Professional and Educational Committee, the Committee of the Society decided in October 1983 to offer Schoolteacher Fellowships for one year as a trial period. They are tenable for one term and can be held in either a University or a Polytechnic. They are intended to enable practising schoolteachers to take part in research and update their knowledge of Biochemistry. Three such Fellowships, value £500 plus £500 for travel and expenses and £500 for research costs, were awarded in 1984. The experiment was considered successful and the Fellowships are being continued. One holder, Mr A. Myers, was invited to contribute to the Education Section of the FEBS Conference in 1986 in Berlin West.

3.12 Awards Committees

In order to co-ordinate the selection of candidates to be awarded the various lectures, medals and fellowships, Awards

Committees are set up annually each with the same nucleus of main Committee members but with additional member(s) representing the endowing organizations. The nucleus consists of the Chairman of the Society, the Honorary General Secretary and three senior ordinary members of the main Committee.

3.13 The Harden Conferences

Sir Arthur Harden, who did so much for the Society in its early stages, willed money to the Society so that in 1967 £4000 [£22,500] became available to the Society to be held in trust. The resulting income was to be used to help defray the cost of publishing the results of original research. At this time the Committee was considering the expansion of the Society's activities by initiating on a modest scale small, informal Conferences similar to the well known Gordon Conferences in the United States. It was a happy thought to acknowledge Harden's great contributions to the Society and to the *Biochemical Journal* by calling these Conferences 'The Harden Conferences'. However, the money from the Harden bequest could not be used to support these Conferences because they were designed for informal discussions with no publication contemplated. The Committee therefore allotted the Harden Bequest to the *Biochemical Journal* and provided money from general funds to support the Harden Conferences. In practice nothing has been disbursed and at the time of writing the Committee is applying to the Charity Commission for a scheme of variation of the objects of the Harden Legacy to allow use of the funds to assist biochemists to attend the Harden Conferences. The Biological Council was called in at an early stage in the discussions aimed at setting up the Harden Conferences so that their scope would not be constricted. Today this ecumenical approach survives in that seven sister societies, members of the Biological Council, are invited to send representatives to the Harden Conference Committee.

The site chosen for the Harden Conferences was Wye College (Fig. 3.21), a residential college of the University of London located in Ashford, Kent. It was an inspired choice and provides a delightfully characteristic English setting for the Conferences, which began in 1969 and are now well established and highly successful. Two are held each year and they are organized and administered by the Harden Conference Committee, which consists of the Honorary Harden Conference Organizer, who is an *ex officio* member of the main Committee and who acts as Chairman, and six members of the main Committee of whom two are the two Honorary Secretaries. As indicated earlier, seven sister societies are invited to send representatives to meetings of the Committee. The pattern of the present Conferences is characterized by

Fig. 3.21. Wye College, University of London: venue for the Harden Conferences.

formal lectures by invited speakers, Poster Sessions and a special lecture by an invited Harden Lecturer. At the moment the Society provides a limited number of bursaries (currently valued at £200) to assist younger members to attend the Conferences. Furthermore six free places are available by competition for each conference; the age limit for these is usually 27.

3.14 Honorary Membership

The institution of this honour was described in Chapter 2. However, by 1958 The Society had no surviving Honorary Members, the last election being that of Sir Charles Martin in 1957. Clearly this aspect of the Society's activities had slid into the background during the immediate post-War rebuilding period. In 1958 the rules of election were amended to "Honorary members shall pay no subscription but shall receive the *Journal* and have all the privileges of membership of the Society including the right to vote". *Transactions* now also comes with the *Journal*. The Society looked at the situation again in 1964 and accepted the recommendation of a sub-committee, in particular that the total number should not exceed 10 and that Honorary Membership should be confined to members of the Society who are of or near retiring age. In March 1984 the total number was increased to 15 and it was decided to donate a plaque to Honorary Members (Fig. 3.22). The names of members elected to the Honorary Membership are given in Table 3.16. The names of a number of distinguished overseas biochemists in this list gives considerable pleasure. A unique aspect of the 500th Meeting of the Society was the attendance of five Honorary Members (Fig. 3.23).

3.15 The Society's Nobel Laureates

Since their establishment Nobel Prizes for Chemistry and for Physiology or Medicine have been awarded with impressive

Table 3.16. Honorary Members of the Society elected between 1944 and 1986

1959	Sir Henry Dale, F.R.S.	1973	A. Neuberger, C.B.E., F.R.S.
	Sir Rudolph Peters, F.R.S.		J. H. Quastel, C.C., F.R.S.
1961	Sir Charles Harington, F.R.S.	1974	Dorothy M. Needham, F.R.S.
1965	Sir John Gaddum, F.R.S.	1979	T. S. Work
1966	R. A. Morton, F.R.S.		Sir Frank Young, F.R.S.
	Sir Charles Dodds, F.R.S.	1982	M. Dixon, F.R.S.
	Sir Robert Robinson, O.M., F.R.S.		E. F. Hartree
1967	Sir Hans Krebs, F.R.S.	1984	F. Sanger, O.M., C.H., F.R.S.
	F. Dickens, F.R.S.	1985	R. R. Porter, C.H., F.R.S.
1969	A. C. Chibnall, F.R.S.		T. W. Goodwin, C.B.E., F.R.S.
	C. R. M. J. de Duve	1986	S. V. Perry, F.R.S.
	W. T. J. Morgan, C.B.E., F.R.S.		R. H. S. Thompson, C.B.E., F.R.S.

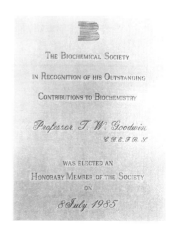

Fig. 3.22. The plaque presented by the Society to Honorary Members on their election.

Fig. 3.23. Honorary Members present at the 500th Anniversary Meeting of the Society (left to right): Professor W. T. J. Morgan, C.B.E., F.R.S.; Professor R. A. Morton, F.R.S.; Sir Charles Harington, F.R.S.; Professor A. C. Chibnall, F.R.S.; Sir Hans Krebs, F.R.S.

regularity to biochemists, and many of the recipients have been members of the Society. The prizes received by members have been distributed almost equally between the two categories (Table 3.17). The list emphasizes not only the strength of British Biochemistry in the post-War period but also, when one realizes that a number of those listed in Table 3.17 are not British, the international character of the Society. A glance at the preceding Tables reveals that the Society's Awards

Table 3.17. Members of the Biochemical Society who have been awarded Nobel Prizes since 1945*

Details for 1911–1942 are given in Table 2.7.

	Chemistry		Physiology or Medicine
1947	Sir Robert Robinson, O.M., F.R.S.	1945	Sir Ernst Chain, F.R.S.
1952	A. J. P. Martin, C.B.E., F.R.S.	1953	Sir Hans Krebs, F.R.S.
	R. L. M. Synge, F.R.S.		F. A. Lipmann, Foreign Mem. R.S.
1957	Lord Todd of Trumpington, O.M., F.R.S.	1958	E. L. Tatum
		1959	S. Ochoa, Foreign Mem. R.S.
1958	F. Sanger, O.M., C.H., C.B.E., F.R.S.		A. Kornberg, Foreign Mem. R.S.
		1962	F. H. C. Crick, F.R.S.
1962	Sir John Kendrew, C.B.E., F.R.S.	1963	Sir Alan Hodgkin, O.M., F.R.S.
1970	L. F. Leloir, Foreign Mem. R.S.	1964	F. Lynen, Foreign Mem. R.S.
1972	S. Moore	1970	U. von Euler
	W. H. Stein	1972	R. R. Porter, C.H., F.R.S.
1978	P. D. Mitchell, F.R.S.	1974	C. de Duve
1980	F. Sanger, O.M., C.H., C.B.E., F.R.S.	1984	D. Milstein, F.R.S.

*A. I. Virtanen (1945), C. F. Cori (1947), A. W. K. Tiselius (1948) and M. F. Perutz (1962) were not members of the Society when they were awarded their Nobel Prizes.

Committee can congratulate themselves in choosing in advance so many future Nobel Laureates for one of the Society's accolades.

One outstanding achievement in this sphere which must be singled out is the award of two Nobel prizes to Dr F. Sanger (Fig. 3.24), in 1958 for his work on the structure of proteins and in 1980 for his work on the structure of nucleic acids. The Society must be proud that most of his protein work was published in the *Biochemical Journal*, but sad that none of his nucleic acid papers appeared there. However, one of the first reviews of this work was contributed to the *Biochemical Society Symposium* No. 30.

3.16 A Royal Charter — To Be or Not To Be?

Early in 1979 the Professional and Educational Sub-Committee (PESC) considered together with the Executive Secretary whether or not the Society should petition for a Royal Charter. The question arose mainly as the result of suggestions that in order to practise in the European Economic Community, biochemists would need some professional accolade such as that provided by the then Royal Institute of Chemistry (now amalgamated with the Chemical Society into the Royal Society of Chemistry). Discussion points raised included (a) the possibility that a Charter would be incompatible with the declared aims of the Society, (b) the Society's responsibility to overseas members, (c) the need to set up an examining body of professional status and (d) the cost of the exercise — probably more than £10,000.

After a great deal of investigation, particularly by the Executive Secretary, and discussion PESC decided in June 1982 that there was no need at present to pursue the possibility of a Royal Charter and this was accepted by the Committee. There is no doubt that in the circumstances prevailing at the time this was a correct and sensible decision, but it did leave the then Executive Secretary rather disappointed that, on the eve of his retirement, he had not achieved a decision which he felt would have moved the Society forwards professionally.

3.17 The Chemical Society Library

The collaboration between the Biochemical Society and the Library of the Chemical Society in the early stages of the Society's development has been succinctly summarized by R. A. Morton, thus:

> "When the Biochemical Society was formed there was no early likelihood that it could have a permanent office or a library. Many members also belonged to the Chemical Society, the Library of which received support from the Chemical Council which

Fig. 3.24. Dr F. Sanger, O.M., C.H., C.B.E., F.R.S. Double Nobel Laureate. Honorary Member, 1984.

negotiated the method by which different societies subscribed to the upkeep of the library.

"The Biochemical Society was invited in 1919 to participate in a scheme whereby the members would be allowed to use the Library in Burlington House under almost exactly the same conditions as Fellows of the Chemical Society. Copies of the *Biochemical Journal* were presented to the Chemical Society and an annual donation made towards the Library. Until this time many requests for an exchange of the *Biochemical Journal* for that of another Society had been made but had not proved feasible. Exchanges now became possible and with the consent of the Library Committee (one member of the Biochemical Society serving on it) exchanges were made with journals that would be a new addition to the Library. A happy and fruitful system of co-operation grew up.

"The Society's subscription was initially £10 [£140] per annum and for some years it remained at that level. Soon after the 1939–1945 War expenditure on books and journals began to rise phenomenally and the Society agreed to take a share in the maintenance of the Library. By 1955 this share slightly exceeded £300 [£2400] and by 1962 it was over £900 [£6200].

"A new basis of assessment was decided upon in 1963. The net maintenance costs of the Library were to be shared on the basis of the membership of the contributing societies with an allowance for overlap which had been calculated in 1961. Until that time the assessment has been calculated on the previous year's costs and not on the current year's costs. This meant that there was always a deficit which had been met by the Chemical Council but whose funds for this purpose were now running out. In 1964 the Society's contribution was £1,080 [£6900] and for the financial year ending December 1968 it was £1,454 [£8100]."

During the last twenty years proposals have periodically arisen in the main Committee that this arrangement need not continue. The changes in the pattern of scientific publications and the increased ease of communication, combined with the small use members make of the Chemical Society Library, have been the main reasons for the suggestion that we withdraw our subscription. This has not happened yet and the *Bulletin* frequently reminds members of the service provided by the Library. The level of our present annual subscription (£1750 in 1984–1985), however, has in no way kept up with inflation since 1968 and obviously reflects a decreased commitment to this particular enterprise, although the Society still has a representative on the Library Committee.

3.18 Archives and the Science Museum

Rather slowly over the past two decades or so biochemists, who are pre-eminently scientists of the present and future, have come to realize that they should look to the origins of their subject. Apparatus or laboratory note-books of the pioneers should not be thrown on the scrap-heap but collected, catalogued and made available for study by future generations of biochemists. The first positive step came when Dr G. B. Ansell was appointed the first Honorary Archivist: on becoming Chairman of the Publications Board Ansell was replaced by Professor G. R. Barker and great progress has been made, including the production of audio and visual tapes of the 'grand old men' of Biochemistry. Some of these, in edited versions, may soon be available on loan to members of the Society.

Many historic pieces of apparatus were being donated to the Society by University Departments and the headquarters at Warwick Court was obviously an unsuitable place for their permanent exhibition. By 1979 the Society had arranged with the Science Museum to establish the 'Biochemical Society Collection' as part of the Wellcome Historical Medical Exhibition to be permanently mounted in the Science Museum. Under this arrangement the historical items in the Society's possession would become the property of the Science Museum although the Society retained full rights of access and exhibition. The early display had not been very satisfactory but a small special exhibition "Sir Hans Krebs (1900–1981). The Discovery of Metabolic Cycles in Biochemistry" was successfully mounted for six months from 1 June 1984. Aspects of his life were illustrated by original archives, photographs, his Nobel Prize and other awards. The apparatus and notebooks concerned with his elucidation of the tricarboxylic acid (Krebs) cycle together with explanatory diagrams helped to emphasize the crucial nature of the discovery to non-scientists. The opening of the exhibition was marked by a lecture by Sir Hans Kornberg (one of Krebs's students) on "The Tricarboxylic Acid Cycle: A Half Century's Retrospect" (Fig. 3.25).

Since then progress has been rapid and it was possible for the Committee to announce in March 1985 that the Science Museum had agreed to house a permanent major biochemical exhibition from December 1986, provided the Society would underwrite a minimum sum. This the Society has done; it has also launched an appeal to Industry and to individuals in an attempt to achieve a target of £250,000, which would allow an entire gallery to be set aside for biochemical exhibits. This target is still (1987) far from being reached. Money was provided by the Society to employ a part-time assistant (H. Kamminga) to provide biochemical expertise for the permanent Museum staff, and it was eventually possible to use the Society's contribution to mount a small exhibition on "Cells, Molecules and Life" to mark the 75th Anniversary of the Society. It was opened on Tuesday 15 December 1986 and should be an important step in developing a permanent section on Biochemistry. The opening was preceded by a lecture by Sir Hans Kornberg and was attended by over 600 senior

Fig. 3.25. Lady Krebs and Sir Hans Kornberg, F.R.S. on the occasion of the opening of the exhibition 'Sir Hans Krebs (1900-1981). The Discovery of Metabolic Cycles in Biochemistry' at the Science Museum in July 1984.

Fig. 3.26. The Society's logos: (*a*) old style; (*b*) new style.

secondary school pupils. It is described in a short article by H. Kamminga [6].

3.19 The Society's Logo

When the Society decided in 1976 to brush up its image it appointed Dr G. A. Snow as a Promotions Organizer, particularly to explore the ways of promoting the Society's publications and to publicize the Society's activities in general. This aspect of the Society's activities will be considered later but arising out of Dr Snow's appointment was his view that the Society's then current symbol, a hexagon surrounding the letters BS (Fig. 3.26*a*), was neither memorable nor particularly appropriate and that a new one was urgently needed. Incidentally it also conflicted with the logo of British Drug Houses. A competition was arranged and members were invited to submit designs for a Society 'logo'. Nearly 100 entries were received and although prizes were awarded, no design was considered suitable, because none exhibited the essential attributes of simplicity and easy recognition. A professional designer was appointed to look into the problem and it emerged that any attempt to symbolize Biochemistry was impracticable and that the two letters BS were difficult to incorporate into an effective design. So it was decided to concentrate on the single letter B and the design eventually chosen (Fig. 3.26*b*) was simple, easy to recognize and suitable for reproduction in a range of sizes. The diminishing white stripes across the face of the letter are intended to give the effect of speedy movement, emphasizing the continuing rapid advance of Biochemistry. As Dr Snow noted "they can also be read as stylized peaks in a chromatogram trace".

This logo now appears on all the Society's official Communications and literature and is apparently satisfactory for this purpose. It remains for individual readers to decide whether or not it is more memorable and less instantly forgettable than the symbol it replaced.

A special logo to celebrate the 75th Anniversary has been designed and appears on the title page of this book.

References

1. Krebs, H. A. (1981) *Reminiscences and Reflections.* pp. 298. Clarendon Press, Oxford.
2. Haslewood, G. A. D. (1984) Manuscript deposited in the Biochemical Society's Archives.
3. Morgan, W. T. J. (1984) Manuscript deposited in the Biochemical Society's Archives.
4. McIlwain, H. (1984) Manuscript deposited in the Biochemical Society's Archives.
5. Morton, R. A. (1969) *The Biochemical Society: its History and Activities 1911–1969.* pp. 160. Biochemical Society, London.
6. Kamminga, H. (1986) Biochemistry on display. *TIBS*, **11**, 343–344.

Chapter 4

The Finances of the Society 1944–1986

4.1 Introduction
4.2 1944–1952 (J. H. Bushill)
4.3 1952–1962 (F. A. Robinson)
4.4 1962–1972 (W. F. J. Cuthbertson)
4.5 1972–1982 (D. F. Elliott)
4.6 Since 1982 (B. Spencer)
4.7 The Costing of the *Biochemical Journal*

4.1 Introduction

Since 1944, five Honorary Treasurers have been in charge of the Society's finances. All have had to face different problems and all have dealt satisfactorily with their particular difficulties, although occasionally activities outside their control made the situation "touch and go"; deft reactions in these cases saved the day. The Society owes a great deal to the Honorary Treasurers, whose hard and effective work has turned the Society into a thriving organization with resources sufficient to allow all the developments outlined in Chapter 3. The reserves they have accumulated should be able to deal with any future contingencies. In short, they have built up a financial structure to support "a forward-looking and virile organization" (J. H. Bushill, Honorary Treasurer, 1944–1954, Fig. 4.1).

The period of office of each of the Honorary Treasurers who served since 1944 covers about ten years and coincides more or less with the different phases of the Society's post-Second World War developments. The rest of the Chapter will thus be sectionalized to consider the developments which occurred under each Honorary Treasurer, although there are obvious overlaps.

4.2 1944–1952 (J. H. Bushill)

Towards the end of the Second World War came the end (1944) of the term of office of that stalwart J. A. Gardner, who had been Honorary Treasurer for 31 years since the establishment of the Society in 1913. His successor was J. H. Bushill, who had served as Assistant Honorary Treasurer under

Fig. 4.1. J. H. Bushill. Honorary Treasurer, 1944–1952.

Gardner from 1941 to 1944. Bushill, who was in office until 1952, saw the Society pass from a period when it was in a condition of delicate financial equilibrium to one in which the increasing number of members and the increasing sale of the *Biochemical Journal* resulted for the first time in the accumulation of reserves, albeit small. This latter development did not please some members, who believed that profit-making was not compatible with the activities of a Learned Society. This as an end in itself is undoubtedly so, but it was also vitally important in order to capitalize the Society so that it would have a buffer against future problems as well as a springboard for developing further activities necessitated by the buoyant development of the science of Biochemistry. This niggling attitude to accumulating surpluses kept recurring over the years but Bushill, in a letter to R. A. Morton, was the first to point out that:

> "When the annual financial reports were presented, it was not unusual for someone to point a finger at the surplus of income over expenditure and say that the accumulation of money was not the function of the Society. Some attempts, which deceived no one, were made at hiding the surplus by transferring money to a 'contingency' account and to justify such action by drawing attention to the continuing rise in the cost of publishing the *Journal*. It was emphasized that, with the increasing size and activities of the Society, paid secretarial and clerical assistance would some day be needed. That was a serious contingency against which, it was stressed, the Society must be prepared."

However, the 'non-profit' lobby won the day over the pricing of the Society Symposia which were introduced in 1948. The remit to introduce such publications contained a statement that the aim should be to market them as cheaply as possible. "It required inspired crystal gazing to decide upon the number to be printed in order that costs and receipts should balance" was the response of a somewhat disillusioned Honorary Treasurer.

To help with the everyday accounting of the Society's finances, Dr Bushill used the services of Mr H. Mears, one of his colleagues in J. Lyons & Co. At the end of his period with the Society, Mr Mears was presented with a gold watch in recognition of his services.

One important act of Bushill was to use a stockbroking firm, Messrs. C. F. Chance & Co., to raise the Society's portfolio of investments.

4.3 1952–1962 (F. A. Robinson)

Dr F. A. Robinson (Fig. 4.2) took over from Dr Bushill in 1952 and served the Society for ten years. The financial situation over this decade was well summed up in a letter (3 July 1968) from Robinson to R. A. Morton:

Fig. 4.2. F. A. Robinson, C.B.E. Honorary Treasurer, 1952–1962.

"During 1951 I had various discussions with J. H. Bushill so that when I took over from him as Treasurer at the A.G.M. in April 1952, I was well briefed as to how things were going. I was also able to obtain the services of an accountant, Mr Mann from Allen & Hanburys [Robinson's employers] and this proved to be very necessary as the amount of money being handled increased rapidly from year to year. In 1951/1952 the finances were not in very good shape and in September 1952, the committee recommended that the subscription be increased to £3.10.0 (£3.50) [£35] per annum and this was argued at a General Meeting held in March 1953; however, A. L. Bacharach (see Chapter 3) and other members criticized the decision to increase the subscription when the Society had such large reserves, but the meeting nevertheless, agreed to the increase. In fact, we only lost 97 members [out of some 2000] as a result of the increased subscription and we had a credit balance of £1,000 [£10,000] at the end of the financial year in 1954. I estimated that had we not increased the subscription, we would have had a deficit of £1,800 [£17,000].

"In 1955, we had a surplus of £2,800 [£25,000] which aroused no comment at the A.G.M. and in 1956 a surplus of £6,000 [£51,000] which was actually greeted with acclamation! In 1957, however, the surplus was only £474 [£3900] as the cost of printing the *Journal* had risen considerably. In fact it cost three times as much as in 1950. In 1958 the surplus was again high, about £4,000 [£31,600] although costs had continued to rise. Each subsequent year ended with a surplus, largely because non-member subscribers paid for the *Journal* by volume and not by an annual subscription. Thus as the *Journal* increased in size, the number of volumes published per annum increased, so that non-member subscribers paid correspondingly more and, in fact, increases in cost were largely met by this increased income from sales. Unfortunately I failed to maintain my regular sum of surpluses and finished with a deficit of £2800 [£20,300] at my last A.G.M. in 1962."

In spite of the difficulties noted by Dr Robinson, the 1950s represented a decade of some stability within the Society when financial affairs in the world at large were anything but stable. However, the deficit in 1962 was a signal of problems to come. The Society's investment policy during this period has been criticized, although on the suggestion of R. H. A. Plimmer (Fig. 1.2), the Honorary Treasurer and the Trustees met Chance Bros in 1954 and the Society's portfolio was reorganized to give an increased income of £120 [£1120] per annum and, at the same time, the Society's money was invested in securities with a better chance of capital appreciation in the long run. However, the investments were still operated under the assumption that the Society was under the control of the Charity Commissioners; that is, investments had to be in gilt-edged securities. That this assumption was unwarranted did not emerge until 1960 when the Trustees sought legal advice on the matter. It transpired that there were no restrictions in the power of investment if it lay with the Committee and not with the Trustees. On the advice of the Society's solicitor an amendment was made to one of the Rules making it clear that Trustees should "deal with the same (i.e. investments) as directed by the Committee". In April 1960, one third of the Society's gilt-edged securities were sold and shares in twenty different industrial equities were bought. The Trustees were then invited to accept responsibility for the investments made on the Committee's behalf. The Trustees felt that their position was now becoming untenable and their reaction to the new situation accelerated the Society's decision to become incorporated (see Chapter 3). This rather late move into equities occurred according to one critic after "the steam had gone out of the stockmarket", that is after the huge increase in the price of equities which occurred in the later 1950s.

4.4 1962–1972 (W. F. J. Cuthbertson)

Dr Cuthbertson (Fig. 4.3) took over from Robinson in 1962, inheriting not only a deficit budget but pressure for increased support for various Society activities; in particular, the *Journal* was still increasing in size. Furthermore, inflation was beginning to be significant and the Society had no permanent home. There was no alternative but to raise the subscription from £3.10.0 (£3.50) to £5 [£36] per annum. At the same time, two new categories of membership were introduced: student membership and joint husband and wife membership. The justification for the student differential was that students, unlike staff, could not claim tax rebate on the membership fee. The joint membership included only one copy of the *Biochemical Journal*; in fact, the realization that the *Journal* was being distributed to members at a loss was probably the

Fig. 4.3. Three Honorary Treasurers (left to right): W. F. J. Cuthbertson, O.B.E., 1962–1972; B. Spencer, since 1981; D. F. Elliott, 1972–1981. (Photographed at a reunion lunch, 23 October 1985, held as part of the 75th Anniversary celebrations.)

turning point which led finally to the Committee's acceptance of the joint membership. The fees for 1963 were set at £2.10.0 (£2.50) [£25.75] for students and £7.10.0 (£7.50) [£53] for husband/wife members.

Meanwhile the Society was anxiously looking for a new home and, as indicated in Chapter 3, the present headquarters was bought for £57,094 [£345,000], which entailed the sale of about 70% of the total investments of the Society. This was a brave decision and, apart from all the advantages which accrued from having a headquarters, the Society benefited significantly from the investment. At the time of writing (1986) 7 Warwick Court is valued at "in excess of £500,000".

By 1966 the Society was in the absurd position that any newly recruited members represented a liability of £3.15.0 (£3.75) [£23.75]. The annual outlay per head for general activities was £4.5.0 (£4.25) [£27.50], the annual cost of the *Biochemical Journal* per head (run-on cost) was £4.10.0 (£4.50) [£28.50] and the annual subscription was £5.00 [£31.95]. In addition the acquisition of headquarters would inevitably incur increased administrative costs. There was no doubt that an increase in the annual subscription was again justified. The Finance Committee recommended an increase to £8.00 [£51.00] and proposed a new concept — membership without the *Biochemical Journal* at £3.10.0 (£3.50) [£22]. This idea was firmly resisted by a group of members on the grounds that it would change the character of the Society. It was the duty, they maintained, of the Society to provide a subsidized *Journal* to each member and, furthermore, the profit made by *Journal* sales to non-members should be used to this end and not to develop new activities which the Officers, in their turn, felt were beneficial to the future well-being of the Society.

Eventually, it was decided to air these problems in the Agenda Papers and the issue for April 1967 printed three statements. The first contained the Committee's views on the future organization of the Society's meetings and publications and advocated the new subscription structure. This was followed by an "opposition" statement drawn up by the three protagonists, J. D. S. Bacon, G. A. Levvy and C. F. Mills. Finally, the third statement gave the Officers' considered answer to the points raised by Dr Bacon and his colleagues.

Another problem closely associated with the change in the subscription pattern was the proposal that Proceedings, the unedited abstracts of Communications to the Society, Symposia lectures etc., which were printed in the *Journal*, should be published separately. The new Proceedings would be distributed with the Agenda Papers and thus give members better value for their basic subscription. This proposed change was also vigorously challenged.

The proposal to introduce the two tier subscription arrangement was passed at the A.G.M. on 13 April 1967; however, the majority was not large enough to carry the motion under the Rules of the Society. In contrast, the proposal to separate Proceedings from the *Journal* was lost. At an Extraordinary General Meeting called at Oxford in July 1967, the subscription change was passed with the necessary three-quarters majority and paragraph 14 of the Articles of Association was appropriately modified. Unfortunately, as the Society had only recently been incorporated (see Chapter 3), there appeared to be some confusion over the arrangements for proxy voting at the E.G.M. Those who had organized the opposition to the subscription proposal felt that if the rules had been applied correctly the three-quarters majority might not have been reached. However, there was no means of knowing whether or not a similar number of 'yes' proxies were also mislaid. The most important outcome of the incident was the realization that voting at the end of a long General Meeting when members are drifting away is not perhaps the best way of settling controversial issues. A ballot of all members is probably preferable.*

The subscriptions finally set for 1967 were £3.10.0 (£3.50) [£21.50] for membership without the *Journal* and £9.00 for comprehensive membership. It was also accepted at the E.G.M. that in future the members' subscription rate for the *Journal* should be brought before the A.G.M. each year and should be less than the run-on cost. The Committee eventually agreed that the price should be "within 70–100% of the run-on cost as assessed by the auditors on the latest figures available". The subscription to the *Journal* for non-members was at this time £27.10.0 (£27.50) [£168].

The problem of the members' subscription to the *Journal* again became acute towards the end of the Cuthbertson decade, mainly because inflation was beginning to bite, as emphasized by considering the "today's prices" quoted in this chapter. By 1970 it was clear that the service cost to each member was higher than the membership fee: £4.9.5 (£4.47) [£22], which would rise to an estimated £5.8.1 (£5.41) [£29] by 1971. It will be remembered that the annual subscription at the time was still £3.50 [£19]. The fact that *Biochemical Society Transactions* was under active discussion at the time (see Chapter 6) made it rather unwise to take a proposal for increased subscriptions to an A.G.M. until 1972; however, this meant that implementation would not be possible until 1973. To add to the difficulty of holding prices steady, the members' subscription rate for the *Journal* was already down by June 1970 to 73% of the run-on cost and was forecast to be below 70% by the end of the year. It would then stand at £8.472 p.a.

---

*A fascinating view of the "politics" of the discussions which eventually led to the changes enumerated in this Section has been provided by Dr J. D. S. Bacon. The manuscript has been deposited in the Society's Archives.

[£46], which would rise to £9.732 [£52.50] if the increased postage rates due to come into force in 1971 were taken into account. On this evidence the members' subscription for the *Journal* was raised to £8 [£36.50] for 1972. Because of the steep rise in the rate of inflation, the 1967 proposal regarding run-on costs would continue to make difficulties so that a proposal was put before the 1973 A.G.M. that the cost of the *Journal* to members should be based on a predicted cost calculated from the run-on cost of the previous year. This amendment was carried and continues in force to the present time (see also Dr Elliott's comments later in this Chapter).

During this decade, the Society became incorporated so budgets had to be prepared in detail and votes adopted which had to be adhered to. As example Table 4.1 shows the summary of votes adopted in 1971 and 1979. The comparison also shows the way the budget was increasing even after making allowance for inflation.

4.5 1972–1982 (D. F. Elliott)

The problems which the Society would face in the 1970s as the result of run-away inflation were signalled by the situation just outlined, which was developing at the end of Dr Cuthbertson's time as Honorary Treasurer and which Dr Elliott (Fig. 4.3, see also Plate 4A) had experienced as Assistant Honorary Treasurer from 1970 to 1971. As Honorary Treasurer he continued mainly in the traditions of his predecessors but had to deal not only with inflation but with unprecedented growth in the Society's activities. He was the first Honorary Treasurer to explain in detail his thinking on the financial problems of the Society at A.G.Ms. Those present at these meetings were confronted with mini-seminars, complete with slides. This successful exercise in communication did a great deal to satisfy the membership of necessary steps which at first sight

Table 4.1. Summaries of expenditure votes for 1970 and 1979

	Vote	1970 (£)	1979 (£)
1.	*Biochemical Journal*	126,889	373,192
2.	Editorial Board	4,980	—
3.	Editorial Office	21,714	104,754
4.	Meetings — General & Travel	12,210	72,835
5.	Other activities	9,107	20,316
6.	Accommodation	3,866	14,150
7.	Administrative Office	27,989	76,462
8.	General Society Expenses	2,285	12,650
9.	Reserve Fund	12,566	—
10.	*Biochemical Society Transactions*	—	55,778
11.	Depot	—	78,580
	Total	221,606*	808,717

*Adjusted to 1979 prices £646,300.

appeared extremely unattractive. Dr Elliott has recorded his thoughts on this period, recollected in the tranquillity of retirement. They are reproduced here and apart from their intrinsic interest, they illuminate many matters discussed elsewhere in this chapter.

"This decade, especially the early part, was one of growth and change as never before experienced by the Society. Main meetings were developing in size and scope, groups were growing apace and creating their own specialist meetings, while the *Journal* was enjoying a period of high popularity as a medium for publication, its size increasing by 25% in the two years from 1970. This very success caused financial pressures which were initially very difficult to control because the need for accurate forecasting and control at the level determined by the budget was not fully appreciated by all concerned with expenditure. The *Journal* was the crucial element in the balancing of the Society's finances. The turnover was so large in comparison with all other sections of business that a small percentage change not forecast had a drastic effect on the overall financial results.

"The central problem of maintaining the financial viability of the *Journal* demanded that all aspects of production and distribution be subjected to very close examination, for in spite of the scientific success, as evidenced by its growing popularity to authors as a medium for publication, sales to external subscribers were actually declining. Sales to members brought no income, rather the reverse. The policy of earlier years whereby members received the *Journal* at the run-on cost determined by the auditors from the accounts of the previous year, resulted in a substantial loss at a time of inflation and expansion at the rates current in the early 1970s. The audited figure became out of date before it could be applied. A loss of £26,000 [£130,000] was incurred in this way during the first three years of the decade. It was necessary, therefore, to ask members at the A.G.M. in 1973 to agree to the proposal that the *Journal* be supplied at a predicted cost based on the run-on cost of the previous year. Sales of the *Journal* to external subscribers continued to fall, albeit slowly, in the face of a determined publicity drive and analyses revealed that although some new customers were being obtained, large institutions were reducing the number of copies taken. It was concluded that the market was saturated and it was the view of the Editorial Board that price increases above a very narrow margin of safety would be counter-productive at a time when the scientific reputation of the *Journal* was growing rapidly. The aim was to keep the *Journal* in front of the widest possible reading public, particularly that overseas. The price was held in 1971 and 1972 but, with growth in pagination over the two years of 25% plus inflation of nearly 10%, together causing heavy increases in the cost of production and distribution, this was sailing very close to the wind from the financial point of view. In fact the balance sheet for 1972 showed an overall loss of £30,271 [£138,000] to the net assets of the Society and thus there was need for a substantial increase in price in 1973 and again in some later years. Increases were kept to the minimum necessary to secure a reasonable financial return but, even so, the combination of growth and inflation, the latter reaching the high point of 24.9% in 1975, resulted in the price of £245 [£335] in 1981. This was 5.4 times the price in 1971 but the size had also increased considerably, as was intended, to the extent of 45% in 1981 after reaching a peak of 54% in 1978.

"After much discussion and argument in every forum of the Society's organization, *Transactions* was ready for launch in 1973 [see Chapter 6]. The size of the new publication turned out to be far beyond predictions, but the time was fortunate because the price of the *Journal* had to be increased by 55% in that year and *Transactions* was offered with the *Journal* at no extra cost for the first year of its life. This ensured maximum publicity for the new publication and perhaps allayed the shock of such a large rise in the price of the *Journal*! *Transactions* was provided free to all members in 1973 and in later years at a predicted cost just as in the case of the *Journal*. *Transactions* was an undoubted success in spite of the adverse effects of mounting printing and postage costs in later years which made necessary some reductions in the amount of space available to authors. Otherwise the price would have been beyond the reach of a large proportion of the membership it was intended to serve.

"The pressure of growth and inflation was also bearing heavily upon the cost of services to members, particularly the printing and postage of Agenda Papers and the organization of meetings. Income coming mostly from investments and sales of the *Journal* had for a number of years provided sufficient funds to keep the membership subscription below the full cost of the services provided, but the gap was widening rapidly and could not indefinitely continue to be filled from this source. It was not considered prudent to raise more income by increasing the price of the *Journal* in this critical period of its life, a question that has already been commented upon. In 1972 it was necessary to ask members for an increase of £2 [£9] on a subscription of £3.50 [£16] and this was agreed at the Annual General Meeting but it proved to be insufficient and a further increase of £1.50 [£6.50] was similarly agreed in the following year. Thus the subscription for 1974 became double that for 1972 and it was not arrived at without considerable protest. Nevertheless, it was still well below the cost of services, as was proved when the *per capita* costs of membership were presented with the accounts in 1975 and could not have been viewed too adversely because the membership actually increased during this period. In parenthesis it should be mentioned that during these difficult years, when the rate of increase in the subscription no doubt gave rise to some alarm, the yearly accounts were presented in considerable detail so that it could be seen how the cost of the essentials needed to maintain a properly functioning Society were soaring year by year. It was possible to hold the subscription at £7 for a further three years, but it was necessary to ask for an increase to £10 [£20] for 1978. Then, for the first time in the history of the Society, it was decided to introduce a differential in subscription for overseas members, few of whom had the opportunity to attend meetings, and a reduction of £2 was made for this category of members. Costs continued to rise relentlessly and an increase to £15 [£23.00] for

1980, again with a reduction of £2 for overseas members, was agreed by the membership. In 1980 it became apparent that finances were heading for a substantial deficit in the following year unless a further large increase in subscription was levied. In view of the fact that meetings accounted for approximately half of the total cost of membership services, it seemed only appropriate that the proportion due to their expenditure should be borne by those who attended meetings. Over the previous years these had undergone unprecedented growth in the scale of facilities offered. Consequently, the Committee proposed that all members should pay a basic subscription of £16 plus optional payment for attendance at meetings, consisting of an annual meetings fee of £8 for home members and £4 for those overseas. The basic subscription would entitle members to receive all other services as before. This proposal was discussed at great length at the Annual General Meeting in 1980 and was carried by a narrow margin. Such unpopular financial measures as those found necessary in 1980 and 1981 were probably responsible for the 10% decline in membership which occurred during these two years, but there had been steady increases in earlier years so that in 1981 the number stood at 102 more than in 1970."

To interrupt Dr Elliott's contribution for a moment; meetings charges were levied for only one year; when the financial position improved dramatically they were quietly dropped but the position was not formally regularized until 1986 when the Committee decided that the meetings charge should not be abandoned but should be zero for the time being — shades of VAT! However, the most recent revenue accounts (see Table 4.2) still show that meetings cost 1.5 times the income from membership fees. Furthermore, the graphs in Fig. 4.4 bear out Dr Elliott's view in showing that increased subscription levels had only a temporary effect on membership numbers. The drop from 1977 to 1981 was reversed in 1982 and by 1985 the number of members had reached an all-time high. Fig. 4.4 further shows that over the period of Dr Elliott's term of office, the subscription merely followed, albeit in a somewhat disjointed manner, the 'indexed' value based on the published general inflation rate.

Table 4.2. Meetings and membership revenue and cost, 1981–1985

Year	Membership income (£)	Meetings costs (£)	Membership costs (£)	Numbers of members	Meetings & membership cost per capita (£)
1981	102,583	59,015	15,590	4537	9.60
1982	70,699	57,416	24,906	5258	15.65
1983	77,531	90,990	35,350	5356	24.00
1984	80,982	102,951	50,635	5662	27.10
1985	92,328	142,546	22,169	6305	26.10

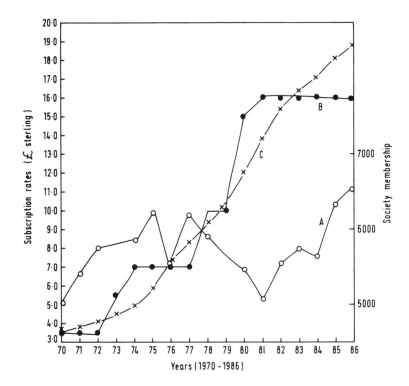

Fig. 4.4. Changes over the period 1970 to 1986 in (A) membership numbers, (B) membership subscription rates, and (C) subscription rates 'indexed' for inflation and normalized to the 1970 rate of £3.50.

"Over the years a great deal of attention has been given to reducing our dependence on sales of the *Journal* by providing more income from other sources. Every vestige of available capital has been invested in stocks yielding a high rate of interest and, from about 1974 onwards, short term investment in the money market of money flowing in from forward subscriptions to publications, has been particularly important, due to the high interest rates available. Although capital gains have not been the objective, such investments nearly always carrying an element of risk not appropriate for a learned Society such as ours, there have been some very useful returns from the sale or redemption of stocks. Particularly noteworthy is the investment in Zambia Copper which cost £18,930 and yielded £63,863 on redemption after only three years. As our export sales of the *Journal* and *Transactions* are mostly to countries whose currency is based on the American dollar, we are vulnerable to changes in the exchange rate if these occur after the dollar price of the *Journal* has been fixed. Fortunately this has only once caused a serious shortfall in income and, on the other hand, in late 1976 a sudden increase in the value of the dollar against the pound resulted in a substantial gain. It is also worthy of note that our publishing activities have made a significant contribution to Britain's exports, the *Journal* and *Transactions* together accounting for approximately £500,000 [£710,000] in 1980. Apart from the *Journal*, *Transactions* has been the most important publication in respect of the income it has provided, whilst other publications and the sale of reprints have also given small returns. The joint venture with the Medical Research Society in the publication of *Clinical Science*, has been successful in that a substantial surplus is now held jointly for the benefit of the two Societies and there have been occasional distributions from this fund to each Society.

"To sum up, the main cause of the financial problems during the ten years from 1970 was rapid growth. Change and growth were necessary to keep pace with the growth of Biochemistry itself but at times enthusiasm stretched our monetary capacity to danger point. The need for strict control in every section of expenditure was the lesson that had taken time to be learned. Inflation was a persistent threat to financial stability; from a figure of 9% in 1971, it reached 24.9% in 1975, then declined somewhat, rising again to 17.2% in 1979, then declining again but by 1981 still remained at 12%. It was undoubtedly responsible for part of the charges borne by the membership."

During this decade the Colchester Depot for storing and distributing the Society's publication was bought (see Chapter 3). Simultaneously with this development, distribution services were offered to other learned societies. This venture was part of a long term plan to make the Society less financially dependent on the profits from the sale of the *Biochemical Journal*. This has developed satisfactorily, apart from a hiccough in the early 1980s, caused by one or two societies withdrawing to develop their own distribution office after learning the 'know how' from the Colchester staff. One or two other societies either withdrew or threatened to withdraw because of what they considered was the excessive commission charged. On investigation, it was agreed that the complaint of high charges was justified and the rate of commission was reduced. This resulted in at least one Society dropping its notice to leave. Recently the administrative activity at Colchester has been streamlined and its efficiency improved. In 1986 the Depot processes distribution of six journals, various supplements and reprint requests. The Membership Office is also now at Colchester. Mrs S. Day (Fig. 4.5) is presently in charge of all these activities.

Efforts over the years and particularly since 1970 to make the Society less reliant on income from the *Biochemical Journal* resulted by 1985 in a small move in the right direction. *Transactions* contributes 4.5% of the total surplus (Table 4.3) and although 'Books' provides less than 1% of the total a change in current policy, which aims at keeping prices as low as possible, could significantly increase the yield under this head. However, further expansion of publications along these lines cannot be expected in the depressed conditions of the publication scene at the moment; indeed, one of the Society's more recent publications (*Essays in Medical Biochemistry*) failed to survive and another (*Bioscience Reports*) has been transferred to a commercial publisher (see also Chapter 6). It is clear that the failure of the Society to venture into a rapid publication journal in the early 1970s when there was a need for such a journal, was financially a lost opportunity. The

Fig. 4.5. Mrs S. Day. Manager of the Society's Colchester Depot.

prestigious and profitable *FEBS Letters* quickly and effectively filled the gap. All these aspects of the Society's publications are dealt with in detail in Chapter 6.

The increased complexity of the Society's finances led to the need for professional day by day supervision of the accounts. A full time accountant (Mr G. Dale, Plate 2A) was appointed in 1980.

4.6 Since 1982 (B. Spencer)

Professor Spencer (Fig. 4.3, Plate 4C) took over from Dr Elliott in 1982 and inherited a stabilized and disciplined situation. Surpluses continue to be made and are currently substantial (see Table 4.3). The investment policy was reviewed and the portfolio of shares has gradually been changing to speed up the disposal of gilt-edged stock and the accumulation of equities. In other words the Society has accepted the advice of the Honorary Secretary to invest for capital growth rather than primarily for income, which, at the time of writing, is satisfactory. This policy, leading to capital accumulation, is an insurance against the time when various technology developments in printing and publishing may render the *Biochemical Journal* in its present form obsolete or, at best, less profitable. Capital assets would be necessary when the time comes, possibly quite soon, for the Society to move to larger premises. [In 1986 this is under active discussion.]

The excellent financial situation in which the Society has found itself in the 1980s has allowed the Committee to approve many new activities (see Chapter 3) under the benevolent but cautious eye of the Honorary Treasurer, who is in a position to relax somewhat the approach of the Society's first Honorary Treasurer, that "the job of a Treasurer is to treasure".

4.7 The Costing of the *Biochemical Journal*

When the membership subscription was separated from the *Journal* subscription the way was open for the Society to exploit its very lively commercial asset — the *Biochemical Journal*. During the 1970s the Committee was very loth to do this mainly because of the deeply held feeling of many that a

Table 4.3. Revenue summaries for the Biochemical Society's operations in 1985

	General Account (£)	Membership (£)	Meetings (£)	*Biochemical Journal* (£)	*Transactions* (£)	Books (£)	*Bioscience Reports* (£)	Distribution (£)	Totals (£)
Total income	349,462	100,324	12.079	1,487,995	148,518	29,303	47,233	87,780	2,262,694
Expenditure	96,506	122,493	154,625	863,808	117,710	23,564	58,275	61,594	1,498,575
Surplus/(Cost)	252,956	(22,169)	(142,546)	624,187	30,808	5,739	(11,042)	26,186	764,119

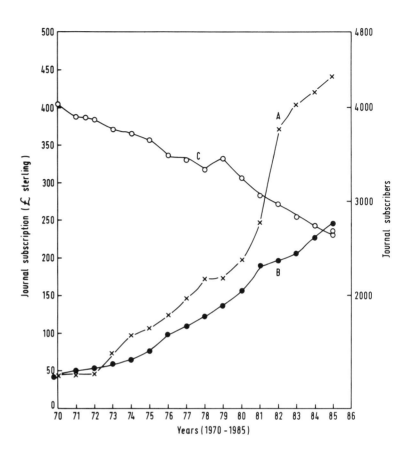

Fig. 4.6. Changes over the period 1970–1985 in (A) *Biochemical Journal* subscription rates, (B) subscription rates 'indexed' for inflation and normalized to the 1970 rate of £45, and (C) subscription numbers.

learned society should not "go commercial", but additionally that making a significant profit might infringe our charitable status and that by increasing the subscription rate we would lose customers. There is no problem with the Charity Commissioners provided the profits are all used to further the science of Biochemistry. One can see that the increases in the *Journal* subscription rate from 1970 until 1980 (Fig. 4.6) were generally restrained. As pointed out earlier, the relatively large increase in 1973 was sweetened by the free distribution to all subscribers of the first volume of *Transactions*. By 1980 the Committee was reconciled to making as much profit out of the *Journal* as the market would stand and a rapid increase in price was implemented from 1981 to 1985; the page cost of the *Journal* in 1980 was 50% of the price for similar journals. 1981–1985 is the only period when the rise in subscription rate has been greater than that of general inflation. If the general inflation index is applied to *Journal* prices (this is, of course, an oversimplification because it is highly likely that changes in the publishing area are different from those in general) and the variation in the size of the *Journal* ignored, then on a 1970 price of £45 p.a. the 1986 price should have been around £2340 p.a. A more realistic view is that on the

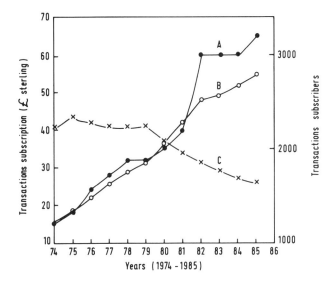

Fig. 4.7. Changes over the period 1974–1986 in (A) subscription rates for *Transactions*, (B) subscription rates 'indexed' for inflation and normalized to the 1974 rate of £15, and (C) number of subscribers to *Transactions* over the period 1974–1985.

basis of a sound commerically judged rate of £445 p.a. the *Journal* had been severely underpriced for the preceding ten years or so and that the 1970 price should have been in the range of £80–£90 p.a.

The question of loss of *Journal* subscribers is a serious one but it is difficult to pin the blame on rising subscription rates. There was already a very steady decline in numbers from 1974 to 1980 (Fig. 4.6) and the rate of decline after 1980, when the subscription prices rose so rapidly, remained the same. The decline is much more likely due to the fall in the size of library budgets associated with the worldwide reduction in science funding by Governments, which has resulted particularly in the cancelling of replicate copies by institutions. The same phenomenon can be observed over the twelve years of the life of *Transactions* (Fig. 4.7), when from 1981 to 1984, a period of constant subscription rate, the number of subscribers continued to fall steadily.

If the raw subscription data (total numbers) in Figs 4.6 and 4.7 are analysed in detail then it emerges that for both the *Biochemical Journal* and *Transactions* the number of U.K. subscribers who are not members of the Society has remained relatively steady over the period 1979–1985. It is the size of the overseas market which has decreased, by 21.75% for the *Journal* and by 24% for *Transactions*. This trend is particularly worrying when it is realized that the overseas market represents 85–90% of the total sales of each journal. The numbers of U.K. members' subscriptions has declined during the same period by 35.4% for the *Journal* and by 51.3% for *Transactions*; the corresponding figures for overseas members are 44.6% and 45.4%, respectively. The relevant data for the *Journal* are given in Table 4.4.

Table 4.4. Distribution of subscribers to the *Biochemical Journal* 1979-1985

Date	Non-Members		Members	
	U.K.	Overseas	U.K.	Overseas
1979	298	3003	254	732
1980	302	2926	240	447
1981	297	2772	216	420
1982	287	2697	184	402
1983	261	2581	180	385
1984	285	2439	183	417
1985	294	2350	164	406

Although every effort is currently being made to improve the circulation of the *Biochemical Journal* (see Chapter 6) saturation point must be close and its high scientific standard must be its most potent weapon in the circulation battle. Librarians cannot resist requests from scientists to subscribe to a journal which is in the top ten of cited biochemical journals.

The prices of the *Biochemical Journal* and *Transactions* to members are now realistically tied to the predicted run-on costs certified by the Society's Accountants as required by the Society's Articles of Association (Article 64) (see also Chapter 3).

So far only the sterling subscription rate has been discussed but North American subscriptions, which are quantitatively the most important source of income, are paid in U.S. dollars and the Society has a U.S. dollar account to deal with these subscriptions. The unexpectedly large variations in the exchange rate between the pound sterling and the dollar added to the difficulties of the Honorary Treasurers in producing sensible estimates, but over the years the situation has probably levelled itself out. Special rates are also in force for Japan, but the rest of the world is tied to the U.K. rate.

Over the period 1970–1985 large and unpredictable variations in the annual number of pages in the *Journal* are apparent. This is an additional hazard in forward planning and is considered in more detail in Chapter 6.

Plate 1

(A) Some past Chairmen of the Main Committee (left to right): T. W. Goodwin, C.B.E., F.R.S. (1971–1974; Honorary Member, 1985); S. V. Perry, F.R.S. (1980–1983; Honorary Member, 1986); T. S. Work (1974–1977; Honorary Member, 1979); F. Dickens, F.R.S. (1966–1967; Honorary Member, 1967; deceased 25 June 1986); G. A. D. Haslewood (1969–1971).

(B) Some previous Honorary Secretaries of the Society. Standing (left to right): P. N. Campbell (1958–1964); H. M. Keir (1970–1977) (currently Chairman of Main Committee); R. H. Burdon (since 1981); D. Robinson (1980–1985); A. P. Mathias (1967–1974); A. N. Davison (1967–1973). Seated: A. C. Chibnall, F.R.S. (1930–1940).

Plate 2

(**A**) Senior Administrative Staff at Warwick Court, November 1985 (left to right): G. W. Dale (Accountant, deceased 1987); Vivienne J. Avery (Assistant Secretary, 1978–1986); G. D. Jones (Executive Secretary); A. G. J. Evans (Editorial Manager); Doris E. Herriott (Meetings Officer).

(**B**) A trio of past Chairmen of the Editorial Board of the *Biochemical Journal* (left to right): A. G. Ogston, F.R.S. (1955–1959); W. V. Thorpe (1959–1963); D. G. Walker (1969–1975).

(**C**) H. F. Bradford (Honorary Secretary, 1974–1981) and G. R. Barker (Honorary Archivist, since 1982).

Plate 3

(A) A. Neuberger, C.B.E., F.R.S. (Chairman of the Editorial Board of the *Biochemical Journal*, 1952–1955; Chairman of Committee, 1967–1969; Honorary Member, 1973) and H. R. V. Arnstein (Honorary Secretary, 1962–1967; Chairman of PESC, 1981–1984).

(B) D. C. Watts (Managing Editor of *Transactions*, since 1977; Chairman of Committee of Management for *Clinical Science*, since 1985) and H. J. Rogers (Chairman of the Editorial Board of the *Biochemical Journal*, 1964–1969).

(C) R. H. S. Thompson, C.B.E. F.R.S. (Honorary Secretary, 1952–1955; Chairman of Committee, 1967–1968; Honorary Member, 1986) and Sir David Cuthbertson (Honorary Secretary, 1945).

Plate 4

(A) D. F. Elliott (Honorary Treasurer, 1970–1981) and W. T. J. Morgan, C.B.E., F.R.S. (Honorary Secretary, 1940–1945; Honorary Member, 1969).

(B) R. M. C. Dawson, F.R.S. (Honorary Publications Secretary, 1973–1980) and R. H. Burdon (Honorary Secretary, since 1981).

(C) B. Spencer (Honorary Treasurer, since 1981) and C. E. Dalgliesh (Honorary Secretary, 1955–1959).

(D) S. V. Perry, F.R.S. (Chairman, 1980–1983; Honorary Member, 1986) and L. Young (Honorary Secretary, 1950–1953).

Chapter 5
The Group Structure of the Society

5.1   Introduction
5.2   Early Developments
5.3   Individual Groups
5.3.1   Molecular Enzymology Group
5.3.2   Pharmacological Biochemistry Group
5.3.3   The Irish Area Section
5.3.4   Neurochemical Group
5.3.5   The Lipid Group
5.3.6   Biochemical Immunology Group
5.3.7   Hormone Group
5.3.8   Techniques Group
5.3.9   The Nucleotide and Nucleic Acid Group
5.3.10 Carbohydrate Group
5.3.11 Industrial Biochemistry and Biotechnology Group
5.3.12 Peptide and Protein Group
5.3.13 Bioenergetics Group
5.3.14 Membrane Group
5.3.15 Regulation in Metabolism Group
5.3.16 Education Group
5.3.17 Monitoring Group Activities

5.1   Introduction

In 1964 the Officers of the Society received a proposal which was to have a greater effect on the future development of the Society than probably any other proposal apart from that to buy the *Biochemical Journal* from Benjamin Moore in 1912. The 1964 proposal was from Dr (now Professor) H. Gutfreund, who requested the establishment of a Molecular Enzymology Group as a "section" of the Society and the provision of £100 [£650] p.a. to run it. As can be imagined, the Committee treated the proposal with as much respect and distrust as if it had been a time bomb. The main worry, which in the circumstances was real enough, was that if the principle were adopted it would splinter the Society irrevocably and that the Society itself would disappear. A second worry, strongly held by some members, was that the "cosy" view of the Society as a Club in which all members knew one another and were

Fig. 5.1. Professor R. R. Porter, C.H., F.R.S. Nobel Laureate. Chairman of the Society Committee, 1977–1980. Honorary Member, 1985.

interested in and could understand each other's work would vanish; at best the result would be an impersonal central administration dealing with isolated groups, rather in the way the MCC runs County Cricket.

To deal with the second point first, however much one sympathized with the protagonists who formulated it, this attitude was entirely unrealistic in the face of the enormous expansion of Biochemistry which was already in full flow in the 1960s and has continued unabated ever since. A central administration has indeed developed which, as we shall see later, deals most effectively with Groups. It is by no means impersonal, it is always making new proposals and exhorting members to suggest new initiatives. If the truth be told it is the Membership which has become impersonal. They take part avidly in the various scientific activities provided but do not take much interest in the running of the Society. At the ballot for Committee Members in 1985 the maximum number of votes cast was 440; the membership numbers about 6500. Furthermore, at the A.G.M. in Oxford in 1985, although there were some 800 biochemists attending the Meeting, only about 50 (probably all over 45 years old) attended the A.G.M., even though one of the most distinguished of contemporary biochemists, the late Professor R. R. Porter, Nobel Laureate (Fig. 5.1), was being elected an Honorary Member of the Society in his home University. Perhaps we should take heart that in the same year 46 members attended the A.G.M. of the Royal Society of Chemistry, a much larger Society than ours. Maybe this lack of interest in "running the Show" is a characteristic of all Societies which provide a satisfactory service.

We can now see that the Group structure was the salvation of the Society; it makes for flexibility of approach to new developments and allows the Society to maintain a major presence in most of these. By far the most frequent comment received by the author from members was in praise of the Group System. Professor Helen Porter (Chairman, 1965–1967, Fig. 3.9) wrote: "the most important thing during my time on the Committee and as Chairman was the introduction of specialist groups about which I held the firm view that if the 'free for all' at every meeting was retained separate and independent groups would arise to meet the needs of the rapidly expanding subject, whereas it was in the interests of all that they should be under a B.S. umbrella ... As I see it, any real contribution to events at the time I made was to fight for separate groups". As a Committee colleague of Professor Porter at that time the author followed with approbation her doughty defence of the Group System against the arguments of some of the Committee "backwoodsmen". Other comments include: "The establishment and development of the Groups System, within the Biochemical Society, did much to keep it together. Those who arranged funds for Groups and who served in Group Committees deserve our warm thanks" (J. K. Grant, Symposium Organizer, 1958–1963). "I personally feel that it was the institution in the 1960s of specialist Groups within the Society that can be considered as one of the most successful adaptations to the vast changes that have occurred in the last generation. These have filled a much felt need which would otherwise, I am sure, have been satisfied with the formation of a plethora of smaller societies" (J. Goddard, Secretary of the Nucleotide and Nucleic Acid Group, 1978–1981). "I have always been an admirer of the Society's Group structure ... during my time with the IUB I have constantly worked to initiate the concept on the international scene" (W. J. Whelan, Honorary Secretary, sometime General Secretary IUB; Fig. 3.10).

5.2 Early Developments

After very thorough, not to say heated, discussions of the Gutfreund proposal, the Committee agreed that a recommendation to form an Enzymology Group be submitted to a General Meeting on 11 December 1964. It is interesting that the adjective 'molecular' was omitted — at the time this was to many members an unacceptable vogue word! The General Meeting, however, approved the formation of a Molecular Enzymology Group; the allegedly perjorative adjective had been restored without anyone apparently noticing it. The rules of the Group were drawn up, approved by the Committee and the first official meeting was held at UCL on 30 April 1965, the subject being a discussion on "The Interaction of Myosin with Adenosine Triphosphate and Actin".

The next proposal to reach the Committee was early in 1966 when they were asked to consider the initiation of a "Pharmacology and Toxicology Group". This was accepted with some apprehension, both financial and scientific, and only with the name of the Group changed to "Pharmacological Biochemistry". The Committee were now clearly soon to be faced with further proposals and needed to formulate a detailed policy for the future. This job was assigned to a subcommittee, which reported strongly in favour of the formation of Subject Groups and suggested guide lines for the formation of Groups and for their financing and administration. These guide lines, which recommended a relaxed but firm central control with a great deal of Group autonomy, were an excellent basis on which to build a successful Group system. As the years have gone by further consideration by a subcommittee (1968) and by Working Parties (1972, 1976) have built on the original guide lines, altering them only to incorporate recommendations for the broadening of the general activities of the Groups. There are currently 16 guide lines, which are worth quoting in full:

1. The total number of Groups supported by the Society will be limited to 15, excluding the Irish Area Section, which represents geographical rather than subject interests.
2. All Groups must have an adequate field of interest and activity. So far as possible the titles of Groups shall be broad enough to allow accommodation of likely growth points in Biochemistry.
3. A signed proposal, from at least 30 people, proposing an organizing committee and defining the field of operation, must be submitted to the Biochemical Society for consideration before the formation of the Group.
4. The proposal must include a draft constitution and a draft programme and budget for the first 12 months of the Group's existence, and must in the first instance be referred to a meeting of the Group Secretaries. Proposals which are endorsed by the latter will be submitted to the Committee of the Society for consideration. The Committee has power to authorize the constitution of Groups, to effect modifications, to refuse the establishment of Groups and to dissolve Groups.
5. Groups may be invited to amalgamate when such a course appears desirable.
6. The need for the continued existence of each Group must be reviewed by the Committee of the Society at least every three years. This is done by consideration of each Group's annual report.
7. The meetings of the Groups will be controlled by their respective organizing committees.
8. Every Group Committee must comprise a Secretary/Treasurer and not more than nine members, with at least one member of the Committee of the Society amongst them, the latter being nominated by the Committee of the Society.
9. Elections to Group Committees will be by postal ballot of Members.
10. Block finance for the Group movement as a whole will be decided by the Committee of the Society, or the Finance Board if delegated with necessary powers.
11. Groups may make small charges to meet the incidental costs of meetings. On occasions of joint meetings arranged with other societies or groups thereof where such other societies have an established practice of levying charges, the Groups concerned may follow the practice of the co-organizers and make similar meetings charges.
12. Groups are expected to give a reasonable account in their annual report of attendance at meetings and, in return for financial support, to submit their accounts.

13. The Society encourages Groups to meet in conjunction with Main Meetings. A grant of £2500 per annum is made to defray the cost of speakers' expenses at such Group Colloquia.
14. Groups will be encouraged to engender interdisciplinary activity and also to initiate proposals for joint Society meetings.
15. The Honorary Meetings Secretary is responsible for the co-ordination of Group activity.
16. The Society expects that Groups will publish the proceedings of at least one Colloquium each year in the *Biochemical Society Transactions*, in extended form of up to 2000 words per paper, and to this end, will pay the publication costs of one Colloquium. Additional Colloquia may be published at the Group's expense. Groups wishing to publish proceedings must give the Society first refusal; an appropriate clause may be found in each Group constitution.

The generous financial support (item 13) is a clear inducement to organize international colloquia and is particularly noteworthy.

In addition there is a general annual subvention of £1500, as well as Committee and Secretarial expenses of £300 and £96, respectively, and an allowance of £200 for entertaining overseas speakers [1986 figures].

The next two Groups to be founded were the Neurochemical Group and the Irish Area Section, which were constituted on 20 September 1967 after being approved at the A.G.M. the previous July in Oxford. Eyebrows were slightly raised at the time at the idea that the Irish Area Section constituted a subject Group, but if it were necessary to have an exception to the rules then no better example could have been found. It represented a most sensible compromise which amicably solved what might have been a difficult situation. This is further discussed under "Irish Area Section" below. Since then there has been a steady stream of new Groups, the last being the Education Group, again not strictly a subject Group. The formation of this Group brought the total up to the maximum currently permitted by the Committee, following the recommendation of the 1976 Working Party (for this purpose the Irish Area Section is not considered a Group). Some hard decisions will have to be made in the future when new proposals come forward, which, if implemented, could result in this number being exceeded. In some cases amalgamation with existing Groups may be possible rather than the drastic step of complete removal of Groups considered "old hat" to make way for newer Groups. However, the pressure for an increase in the number of Groups may become irresistible. A recent sensible extension of a Group's activity was implied by the change of name in 1982 of the Industrial Biochemistry Group to the "Industrial Biochemistry and Biotechnology Group".

A complete list of the present Groups is recorded in Table 5.1, together with the names of the founding Secretaries and Chairmen, whose enthusiasm probably were responsible for bringing the Groups into being. A full up-to-date list of Group Committee members is recorded annually in the Society's Annual Reports.

5.3 Individual Groups

5.3.1 Molecular Enzymology Group

This Group, having been first in the field, came of age in 1985, and like all good offspring, has matured most responsibly and effectively and is now a pillar of the Society. One cannot even detect any adolescent hiccoughs in its steady development.

5.3.2 Pharmacological Biochemistry Group

In the decade between 1940 and 1950 a small group of specialist biochemists were concerning themselves with the metabolism of drugs, pesticides, herbicides and similar foreign compounds. British workers were well to the fore in this development, none more so than R. T. Williams, Professor of

Table 5.1. The Society's Groups with their first Secretaries and Chairmen

Group	Date of Founding	First Secretary	First Chairman
Molecular Enzymology	11 December 1964	A. P. Mathias	B. R. Rabin
Pharmacological Biochemistry	6 July 1966	D. V. Parke	T. J. Franklin
Neurochemical	20 September 1967	H. S. Bachelard	G. B. Ansell
Irish Area Section	20 September 1967	W. K. Downey	D. T. Elmore
Lipid	14 February 1968	C. H. S. Hitchcock	T. W. Goodwin
Biochemical Immunology[1]	18 April 1968	D. R. Stanworth	R. R. Porter
Hormone[2]	19 February 1969	V. H. T. James	G. A. D. Haslewood
Techniques[3]	16 April 1969	G. N. Graham	J. H. Ottaway
Nucleotide & Nucleic Acid[4]	8 April 1970	D. W. Hutchinson	A. S. Jones
Carbohydrate	15 October 1970	A. R. Archibald	J. Baddiley
Industrial Biochemistry and Biotechnology	17 December 1970	E. F. Annison	P. J. Heald
Peptide & Protein[4]	18 February 1972	R. C. Sheppard	H. N. Rydon
Bioenergetics	4 July 1972	D. E. Griffiths	F. R. Whatley
Membrane	13 April 1973	A. H. Maddy	J. A. Lucy
Regulation in Metabolism	7 July 1977	J. Mowbray	D. A. Hems
Education	20 July 1984	T. G. Vickers	E. J. Wood

[1] Jointly with the British Society of Immunology.
[2] Jointly with the Society of Endocrinology.
[3] Jointly with the British Biophysical Society.
[4] Jointly with the Chemical Society (Royal Society of Chemistry). Recently renamed Nucleic Acid and Molecular Biology Group.

Biochemistry at St Mary's (Fig. 3.16), and it was fitting that one of his group, Dr D. V. Parke (now Professor at the University of Surrey), should propose in February 1966 the founding of a "Biochemical and Pharmacology and Toxicology Group". The first reaction of the Committee was equivocal; the possible financial commitment worried some members. However, in May 1966 the proposal was accepted but as stated earlier in this chapter the Committee insisted on a change of title to that which still holds today. As indicated in the comments of Dr P. T. Nowell (Secretary of the Group) below, some of the steam had gone out of the Group by the mid-1970s and the 1976 Working Party recommended that the Group be merged with the Industrial Biochemistry Group. Eventually this suggestion was not implemented and the Group still survives and continues to make important contributions. Dr Nowell assesses its impact over the years:

"The formation of the Pharmacological Biochemistry group was a progressive and enlightened move which had far-reaching repercussions. Although its original proponents were biochemists, it brought together a wide variety of scientists concerned with pharmacology and toxicology, including clinicians, pharmaceutical chemists, histopathologists and immunologists from both academic and industrial establishments on an informal basis. At the time, there was virtually no other forum in the U.K. where this could occur, since the other main societies involved with pharmacology, notably the British Pharmacological Society and the Physiological Society, tended to be more restrictive in their activities with emphasis being concentrated primarily on pharmacodynamics and the electrophysiological aspects of pharmacology.

"For approximately 10 years from 1966 to 1975, the Pharmacological Biochemistry group occupied a key position in bringing biochemistry, pharmacology and toxicology into close proximity with each other by giving close attention to molecular mechanisms and their wider implications. Following its success, other multi-disciplinary groups emerged under different auspices, with usually more expanded or specialized functions. The most notable of these developments were the formation of the clinical pharmacology section of the British Pharmacological Society and the independent drug metabolism group, together with the toxicology club and the drug metabolism group; the impetus [for the formation of these groups] was from biochemists who saw the requirements for these in the light of international events. They were quickly joined by other scientists, particularly from the growing band of those working in these areas in industry.

"Despite all the above happening, the Pharmacological Biochemistry group continued to function, although perhaps not with quite such a wide range of activities as previously. In addition, other Society groups such as the industrial biochemistry group and the neurochemical group frequently became involved with pharmacological and toxicological topics. The Society in fostering these activities has been a major influence in contributing to knowledge about the actions of drugs and toxic agents. Although this type of work can often become very specialized because of the type of procedures used, it must of necessity proceed on a broad front along multidisciplinary lines in order to give a meaningful overall picture".

5.3.3 The Irish Area Section

In 1964 E. R. Tully and L. Downey organized a Christmas reunion of graduates of the Biochemistry Department of University College Cork at which papers were read by a number of returning alumni. At that meeting an informal Working Party was set up to consider the desirability of establishing an Irish Biochemical Society. This was at the time when the Society was beginning to develop its Group structure and, with positive support from the Society's Officers and Chairman, the concept of an Irish Area Section within the parent Society emerged. In 1966 a meeting of more than 200 Irish biochemists decided not to form a separate Society but agreed to ask the Society to authorize the formation of an Irish Section with which was coupled the request to hold one Society meeting in Ireland per year. At that time an official meeting in Ireland was held only every sixth year. On 13 July 1967 the General Meeting of the Society accepted both proposals and the Section was formally inaugurated on 20 September 1967 with L. Downey, a protagonist in the early negotiations, as its first Honorary Secretary. The Society's meetings are now held in rotation at the three constituent colleges of the National University of Ireland, at Trinity College Dublin and at Queens University, Belfast.

The successful conception and parturition of the Section were due not only to the enthusiasm of the local activists but also to the far-sightedness of the Society's Officers at that time in appreciating and encouraging the natural aspirations of Irish biochemists to have their own formal organization.

The Section has maintained the liveliness of its early years and has made many innovations, in particular "The Irish Lecture Tour". Annually a distinguished biochemist is invited to lecture at the four major University centres during a four-day whistle-stop tour. The Section's Annual Special Meeting for predoctoral students has also been a very successful development.

The continuing success of the Section reflects the hard work of the local Officers and Committee over the years. Professor M. G. Harrington (U.C. Dublin), who has provided much information about the Section, claims that the success has much to do with the 'simple organization' of the Section Committee. In the early days "the Section Committee was set up annually by a gentleman's agreement. Part of the unwritten agreement was the exclusion therefrom of those over 35. The elder statesman element was provided by the nominated representatives of the Society Committee". This relaxed approach was encouraged by a "Guinness Lunch", which was kindly provided at one of the three annual Committee meetings. Apparently, occasional well-meaning attempts to improve the efficiency of one in three Committee meetings by eliminating the "Guinness Lunch" "have been singularly unsuccessful". Apart from the provision of these legendary lunches the Guinness Research Laboratories, through the good offices of Dr A. K. Mills, the Research Director at that time, helped in many other ways. Dr Mills arranged facilities for Committee meetings, provided financial support and actively encouraged his younger colleagues to take a positive part in the business of the Section; Dr R. Letters, for example, was Secretary for some years.

5.3.4 Neurochemical Group

This was the third Group to be established, coming into formal existence on 20 September 1967, and it has had a distinguished history. Professor H. Bachelard (St Thomas's Hospital Medical School) has kindly provided a detailed history of the Group in its relation to the development of the International Society for Neurochemistry and a European Society of Neurochemistry (ESN). It is reproduced here with only minor amendments and omissions:

> "Neurochemistry has formed an integral part of the interest of chemists and biochemists since the time biochemistry was first recognised as a distinct scientific discipline, so any appraisal of the development of neurochemistry in the U.K. should include an acknowledgement of the early contributions of some of our eminent biochemists. In addition to the pioneering chemical analyses of the brain, performed by Thudichum over a century ago (below), many biochemists found in the brain their major research interest.
>
> "One of the first specifically biochemical posts in the U.K. was that of Sydney A. Mann, appointed in 1901 to the Central Pathological Laboratories of the London County Council's Mental Health Services. Mann was a founder member of the Biochemical Society, and many of his publications reflected his interests in cerebral and endocrinological themes. He was prominent among those who contributed to the development of neurochemistry as a distinct speciality within mainstream biochemistry and cognate to the neurosciences as well as to psychiatry.
>
> "Notable amongst these pioneers in the years between the two World Wars was Sir Rudolph Peters (Fig. 3.11) who used cerebral preparations in his classical work on vitamins at Cambridge. Also at Cambridge, and subsequently in Cardiff, Judah Quastel (Fig. 5.2) was performing his innovative work on the metabolism of acetylcholine and the monoamines, and also on barbiturates and

Fig. 5.2. Professor J. H. Quastel, C.H., F.R.S. Honorary Member, 1973.

anaesthetics. Derek Richter in the late 1930s with Hermann Blaschko in Cambridge, did much towards characterizing the monoamine oxidases; Richter was subsequently at Mann's L.C.C. laboratories and then in Cardiff, from where many papers on amphetamines and catecholamines emerged. A major proportion of the scientific announcements of the work of all these scientists appeared in the *Biochemical Journal*.

"In the late 1940s and early 1950s, neurochemical themes became prominent as parts of organized meetings of the Biochemical Society, as reflected in the Society's Symposium on "Metabolism and Function of the Nervous System" in 1952. This was organized by Henry McIlwain, another of the major contributors to the early development of the subject. He was at that time at the Institute of Psychiatry in London — an institution which is, interestingly, a linear descendant of the Central Pathological Laboratories of the L.C.C. attached to the Maudsley Hospital. Neurochemistry has frequently formed a vital part of subsequent meetings of the Biochemical Society in many parts of the country.

"Concurrently with these developments, neurochemistry was becoming recognized and organized at international level. Many members of the Biochemical Society contributed to International Neurochemical Symposia (the fore-runners of the International Society for Neurochemistry) between 1955 and 1965; among them, Hermann Blaschko, Henry McIlwain (who has recently written on the early days of the ESN) and Derek Richter were on the organizing committees. These Symposia, like the meetings of the International Society which succeeded them, were held only every two years — occasions were therefore sought for smaller and more frequent meetings in Britain. As a result of correspondence between Brian Ansell and Henry McIlwain around 1960, the idea of a national neurochemical group or club began to be formulated. With the announcement of the first proposed group within the Biochemical Society (the Molecular Enzymology Group) this structure was seen as a welcome framework for neurochemists. Henry McIlwain and Herman Bachelard then contacted interested biochemists early in 1967, who met informally in May 1967. It was agreed that the Biochemical Society be asked to approve the formation of this Group, and that Herman Bachelard would attend to the details as provisional Secretary. Official approval was granted during the Oxford meeting in July 1967 and the first scientific meeting of the group took place at the Institute of Neurology, London, in November 1967. Over the first three full years of operation, four meetings were held each year with average attendances of $c.$ 100. During this period a policy was designed to render the A.G.M. attractive — by offering refreshments and having an historical talk; speakers included J. N. Cumings. D. Richter, Dr R. Peters and H. McIlwain. Since then this momentum has been maintained.

"In 1969 and again in 1970, the possibility of a European Society for Neurochemistry (ESN) was mooted but not formally initiated. Finally, largely as a result of initiatives from the Neurochemical Group through Alan Davison, the ESN was established in 1976. The first ESN Executive Committee to be elected, 1976, included four members of the Group Committee and the first formal meeting of the Society was held in Bath in 1976. The organizing committee were all members of the Biochemical Society.

"Special Workshops (roughly biennial) were initiated in 1972 (to get clinicians and scientists together on specified topics). These have all been published as having emanated from the Neurochemical Group.

"Thudichum Medal Lectures were inaugurated in 1974, to honour eminent scientists who had made outstanding contributions to neurochemistry and related subjects. Although Thudichum (Fig. 5.3) was an undoubted pioneer of brain chemistry a century ago, his contribution to the overall academic development of the subject has been controversial [1], so there was some doubt expressed about the wisdom of striking a Medal in his honour. Nevertheless the majority view prevailed and the attractive Medals were struck in hall-marked sterling silver in a batch of 11 (to save money!) (Fig. 5.4). The dies (the most expensive items) are stored in the Biochemical Society safe for future use. (The cost of preparing and striking the Medal came from Group funds.) The lectures have become very happy occasions; recipients of the Medal so far have been: H. Blaschko (1974), H. McIlwain (1975), M. Vogt (1976), H. Kosterlitz (1980), V. P. Whittaker (1983). (Four of the five lectures have been published in *Transactions*.)"

Fig. 5.3. Professor J. L. W. Thudichum (1829–1901).

5.3.5 The Lipid Group

A meeting of 53 members interested in lipids was arranged at the Unilever Research Laboratory, Colworth House in June 1967; four papers were read and a temporary Committee was set up to put forward plans to the Society for the formation of a Lipid Group. These were accepted and the Group came into being on 14 February 1968. Dr A. T. James of Unilever provided considerable support in these early stages and has continued to help over the years.

5.3.6 Biochemical Immunology Group

This began as the Immunoglobulin Discussion Group thanks to the persistence of Dr D. R. Stanworth, who eventually became its first secretary, and the encouragement of the Society, whose sub-committee on Groups (1966) had suggested immunology as an area for development. In spite of lukewarm support in the early stages from two eminent biochemical immunologists (one with sublime lack of logic, whilst apologizing for the delay in answering Stanworth's letter because he had been in the U.S., felt that the formation of a Group might entail "a considerable amount of travelling"), a draft constitution and proposals for Committee membership were accepted by the Society on 18 April 1968. The first formal scientific meeting was held at the Institute of Child Health on Friday, 7 June 1968 with the late Professor R. R.

Fig. 5.4. The Thudichum Medal.

Porter (Fig. 5.1) in the Chair. The British Society for Immunology helped financially in the first year with a contribution of £15 [£90], which was offered without obligation as a token of their interest. Eventually, on the recommendation of the 1976 Working Party, the Discussion Group evolved into the Biochemical Immunology Group sponsored jointly by the Biochemical Society and the British Society for Immunology. The new name adequately mirrored the decision to widen the subject coverage from immunoglobulins to all biochemical aspects of immunology.

5.3.7 Hormone Group

This Group, which started life early in 1969 as the Steroid Biochemistry Group, was transformed into the Hormone Group on the recommendation of the 1976 Working Party, which also recommended that it should become a joint Group sponsored by the Biochemical Society and the Society for Endocrinology. This change also took place.

5.3.8 Techniques Group

The precursor of the present Group, a joint Group of the Society and the British Biophysical Society, was the Computer and Instrumentation Group formally constituted in April 1969 after a preliminary meeting in 1968. It was therefore the first joint Group approved by the Society. The 1976 Working Party's recommendations that it should continue as a jointly sponsored Group with the Biophysical Society and that it be renamed the Techniques Group were implemented in 1978.

5.3.9 The Nucleotide and Nucleic Acid Group

The origin of this Group differs from that of other Groups, except the Protein Group (q.v.), in that the initiative was taken by the Chemical Society (now the Royal Society of Chemistry), which formed a Nucleotide Group to "encourage the discussion of the chemistry including the biological chemistry of nucleotides, nucleosides and nucleic acids". The first meeting of the Group was held in Birmingham on 9 January 1968. However, it soon became clear to Dr R. T. Walker (Birmingham), the driving force in the formation of the C.S. Group, and to Professor G. R. Barker (Manchester, currently Honorary Archivist, Plate 2C) that pressure was arising within the biochemical community for the formation of a similar Group. Together they eventually persuaded the two societies to found the Joint Nucleotide Group in 1970, and thus the nonsense of the existence of two competing Groups was avoided. It is fair to say that the enthusiastic support given by the Biochemical Society has allowed the Group to blossom, whereas the R.S.C., according to one member, barely tolerated the Group because it was a possible threat to the chemical establishment. Be that true or not the financial contribution of the Society to the Group is considerably greater than that of the R.S.C. Following the Working Party recommendation in 1976, it was renamed the Nucleotide and Nucleic Acid Group after some heart-searching from the Group Committee.

It is appropriate here to consider the suggestion made by the Working Party and approved by the Society Committee that as Biochemical Genetics and Protein Biosynthesis were under-represented in the Group structure an application to form a Group in this important growth area would be welcomed. This was promptly taken up by Professor P. N. Campbell (Plate 1B), who suggested a Group on "Gene Expression and Protein Synthesis". The Joint Nucleotide and Nucleic Acid Group Committee reacted unfavourably to this idea, claiming that their programmes covered this subject and an inevitable and unacceptable overlap would occur and that according to a letter from Professor G. R. Barker, the then Chairman, to Professor Campbell, "there is much flexibility in the present Group, whatever the name may be, and that there is no problem in providing for the needs you mention through better communication between the Group Secretary and his customers"; and there the matter rested. There is no doubt that the case made by Professor Barker at that time was correct but such is the appeal and magnetism of fashionable words that many observers of the Society's activities feel that Molecular Biology is not effectively catered for. For example, Professor W. J. Whelan commented in a letter (now in the Society's Archives) to the author, which is generally appreciative of the Group System: "If I look at the Biochemical Society's Groups, it is to see that genetics and developmental biology are conspicuous by their absence. I do believe that it is up to any organized group of biochemists to welcome and encourage the growth of exposition, discussion, debate and publication on these new areas within the Society structure itself. The kind of new Groups to which I refer might well be organized in conjunction with other societies, as is the case for five of the Society's Groups".

In what appears to be a reasonable compromise in the face of mounting pressure the Committee recently accepted the recommendation that the Group be renamed the "Nucleic Acid and Molecular Biology Group". This has now (1987) been officially approved by the Committee of Group Secretaries.

5.3.10 Carbohydrate Group

This Group came into being on 15 October 1970 after groundwork by Professor Walter Morgan (Plate 4A) and

Professor (later Sir James) Baddiley, the latter being the first Chairman.

5.3.11 Industrial Biochemistry and Biotechnology Group

The Industrial Biochemistry Group, formed on 17 December 1970, fared rather less well than other Groups and only just survived the hatchet when the 1976 Working Party discussed its future and recommended a merger with the Pharmacological Biochemistry Group. However, it did survive and in 1982, because of the rapid advances in genetic engineering which have such significant industrial possibilities, it was renamed the Industrial Biochemistry and Biotechnology Group. Its meetings have "a strong professional emphasis as well as the usual academic content".

There are two organizations, supported by the Society, which impinge on the activities of the Industrial Biochemistry and Biotechnology Group. The British Co-ordinating Committee for Biotechnology (BCCB) was formed by a group of interested parties, including the Biochemical Society, meeting at the Society of Chemical Industry: its first objective was to organize the second Congress of Biotechnology in Eastbourne in April 1981. Its long term aims are, in summary (i) to provide a forum for British Societies to exchange views and decide on concerted action; (ii) to advance the science and technology of Biotechnology; (iii) to assist members in co-ordinating meetings; (iv) to provide a focal point of references with Government Departments and other similar organizations and (v) to co-ordinate and safeguard British interests within the European Federation of Biotechnology (EFB).

EFB was established in September 1978 during a Biotechnological Congress at Interlaken in which the Biochemical Society was one of 35 European Scientific Societies taking part. The objective of the Federation, which is a voluntary and non-profit-making organization, is to advance Biotechnology as an interdisciplinary field of research and to further the application of such advances to manufacturers' processes. Up to the present its main activity in moving towards these goals has been to establish working parties to survey and report on certain areas of Biotechnology. Reports of such working parties are routed to the Society via the BCCB (the agreed procedure between EFB and BCCB) and thus to the Industrial Biochemistry and Biotechnology Group.

The Society nominates appropriate representatives to the General Assembly of EFB and pays their expenses. BCCB makes recommendations to the Society for nominations to working parties and other activities of EFB; these the Society can either accept or reject as it chooses.

Good exploratory work is being achieved by EFB and BCCB but some improvements in liaison with the Society will occur when a few administrative rough edges are filed smooth.

5.3.12 Peptide and Protein Group

The formation of a Protein Group was approved by the Council of the Chemical Society on 3 April 1968, but the close links with Biochemistry were soon apparent. By 1970, informal discussions were proceeding with the Biochemical Society about the possibility of the setting up of a joint Group and these were formalized on 22 October 1971 by a letter from Dr R. C. Sheppard to the Executive Secretary:

> "I write on behalf of the Chemical Society Protein Group. For some time past the Committee have been considering the desirability of a formal association with the Biochemical Society, and I now write to suggest that the Group becomes a joint Group of the two Societies.
>
> "The Protein Group was founded in 1968 to provide a forum for discussion between scientists of all disciplines with interests in peptides and proteins. Membership has grown rapidly and now stands at 332. Of these, only 171 are Fellows of the Chemical Society, and I believe that a large proportion of the remainder, as well as many of the Fellows, are members of the Biochemical Society. Four of the five members of the present Committee are members of both the Societies. Of the eight meetings held by the Group, two have been held jointly with the Biochemical Society. There thus exists already a close relationship of the Protein Group with both Societies.
>
> "There should, of course, be no element of competition between the Protein Group and any existing Group of your Society. The interests of the Protein Group are very broad, and individual meetings often cover a wide range of topics. If an occasional overlap with the interests of another more narrowly based Group should occur, we would envisage that the particular meeting should be held jointly with the other Group concerned. In this connection, it is worth noting that one of our Committee, Dr R. Perham, is also a Member of the Committee of the Molecular Enzymology Group. Arrangements such as this should ensure that no difficulties arise.
>
> "I understand that the Nucleotide Group is now a joint Group of the two Societies. If the Biochemical Society is agreeable, we would be happy to accept a constitution essentially identical to that of the Nucleotide Group."

This proposal was received with enthusiasm by the Biochemical Society and the Joint Group was formally set up on 18 February 1972.

The field of interest in this Group, which could be almost the whole of Biochemistry, is generally accepted as peptide and protein structure.

5.3.13 Bioenergetics Group

On 4 July 1972, a Bioenergetic Organelle Group was formed and functioned as such until 1978 when its name was changed to the Bioenergetics Group following the recommendation of the 1976 Working Party. This has close ties with the IUB/IUPAC Bioenergetics Group, which was formed after some effort by Professor W. J. Whelan, lately Secretary General of IUB, who is an admirer of our Group structure. His further efforts, "likened to pulling teeth", have now resulted in the formation of seven IUB Groups, some, like the Bioenergetics Group, co-sponsored by other Unions. However, at the moment of writing no other Society Group has formal ties with the IUB Groups.

5.3.14 Membrane Group

Formed in 13 April 1973, the Membrane Group continues to serve an important need in providing a forum for experts in this increasingly influential aspect of Biochemistry.

5.3.15 Regulation in Metabolism Group

The 1976 Working Party recommended that one new Group should be initiated to cover the area of metabolic regulation. As a result of this recommendation the Regulation in Metabolism Group was founded on 7 July 1977. Thus, after a spate of new Groups in the late 1960s and early 1970s, four years had elapsed between the formation of the Membrane Group and this Group, the last scientific Group to come into existence.

5.3.16 Education Group

This Group was set up as recently as 1984 as a result of the concern that the proper training of biochemists is becoming more and more important as knowledge and specialization increase at an alarming rate. It is now accepted that education of biochemists is a legitimate activity of the Society, although this view has not always been accepted, particularly in the 1960s. Before the current upsurge in interest in biochemical teaching the Society held a meeting in the very early days on the teaching of medical students, and more recently two Colloquia on the training of biochemists; the last two were held on 13 July 1961 in Oxford and on 15 September 1966 at Aberystwyth, chaired by the late Professor K. S. Dodgson and Professor G. R. Barker, respectively. The proceedings of both Colloquia were published. In 1967 the Society submitted a memorandum to the Royal Commission on Medical Education, reproduced in the Annual Report for 1967. The establishment of the Education Group, the ultimate accolade of Society respectability, was the result of the initiative of Dr E. J. Wood, who organized a half-day discussion session and an 'education corner' in the Poster Session during the Society's meeting at Leeds 18–20 July 1984. The interest aroused made it possible to collect the 30 signatures required before the Committee will consider the formation of a new Group. The main aims of the Education Group are (i) to hold colloquia and present Posters and demonstrations on educational topics at Society Meetings and (ii) to facilitate exchange of educational technology — video tapes, computer-assisted programs etc.

Further aspects of the Society's present positive policy on Education are discussed in Chapter 7.

In a different way from the Irish Section, this Group is also not a conventional subject Group and assessment of its impact or otherwise is for the future to decide.

5.3.17 Monitoring of Group Activities

The overall activity of the Groups is monitored by having one member of the General Committee nominated as a member of each Group Committee. The Group Secretaries meet once a year to co-ordinate activities and discuss future developments.

Reference

1. Drabkin, D. L. (1958) *Thudichum: Chemist of the Brain*, University of Pennsylvania Press, Philadelphia.

Chapter 6

The Society's Publications

6.1 Introduction
6.2 The *Biochemical Journal* 1945–1965
6.3 The *Biochemical Journal* 1965–1986
6.4 *Clinical Science*
6.5 *Biochemical Society Transactions*
6.6 *Bioscience Reports*
6.7 *Essays in Biochemistry*
6.8 *Essays in Medical Biochemistry*
6.9 Biochemical Society Symposia
6.10 *Biochemical Society Bulletin*
6.11 Special (Occasional) Publications
6.12 The Future

6.1 Introduction

In Chapter 3 the post-War developments in the Society were described but publications were deliberately left for consideration in a separate chapter. It is obvious from the early History of the Society that the acquisition of the *Biochemical Journal* in 1912 and its development into a leading international outlet for biochemical papers had by 1944 set the Society on a reasonably firm financial foundation (see Chapter 4) on which the present impressive edifice has been built. Although, as will become clear later in the chapter, periods of friction sometimes occurred between the General Committee and the Editorial Board, the general impression is that of an efficiently run journal whose Editors have reacted responsibly to the real difficulties which have been thrown up by the General Committee.

Although the *Biochemical Journal* is the flagship of the Society's publication fleet, it is extremely well supported by *Transactions*, which turned out to be the very opposite of the destroyer predicted by some conservative members of the Society. Indeed in its own sphere it has quickly developed in a way of which the Society can be justly proud. *Clinical Science* is a successful joint venture with the Medical Research Society. *Essays in Biochemistry* made its mark some 20 years ago as an annual publication and the series *Biochemical Society*

Symposia is also well established. Not all the publishing ventures have been successful; *Essays in Medical Biochemistry* closed after four volumes and *Bioscience Reports* has survived by being transferred to a commercial publisher. Also, occasionally opportunities have been lost. All these topics will be enlarged upon as the chapter progresses.

6.2 The *Biochemical Journal* 1945-1965

The retirement of Arthur Harden in 1937 as Editor after 25 years was clearly the end of an era (Chapter 2) but the Society was lucky in that C. R. (afterwards Sir Charles) Harington (Fig. 2.7), who had been Harden's assistant for seven years, accepted the invitation to fill Harden's post and served as senior Editor until 1942. During this period he was helped by three associate Editors, S. J. Cowell, F. Dickens (Plate 1A) and F. J. W. Roughton.

Harington, when he resigned on being appointed Director of the National Institute for Medical Research, recorded his views on this period in Morton's *History* [1]:

> "As it happened I welcomed the invitation, little realizing what I was letting myself in for, because at that particular juncture I had no serious responsibilities outside my own research and I was anxious for a task that required a different type of effort; this I certainly got. I had had, of course, no previous experience of editorial work and my appointment was an indication of the somewhat light-hearted view that the Committee at that time took of the duties required of the editors of the *Journal*. I am sure that the very thought of a professional editor would have filled them with horror.
>
> "By the time I joined Harden he had trained himself to be an excellent editor. He possessed an equable temperament, could work rapidly with economy of effort and was an admirable colleague for whom I had a great respect, which increased as time went on. I soon learnt, however, that he expected his co-editor to possess the same capacity for getting through the work as he himself had acquired; no sooner had I been appointed than he told me that he had arranged his summer holiday for certain dates which would mean that he would have to leave me to prepare the next number of the *Journal* for press by myself. I neither relished the prospect, which was somewhat alarming, nor enjoyed the performance — especially as this involved the almost complete re-writing of one of the papers — but there is no doubt that this drastic introduction did give me a measure of confidence (perhaps too much) and taught me in three weeks of hard work what I might otherwise have taken a long time to learn.
>
> "In the early and amateurish period of which I am writing, editorial practice was admittedly dictatorial. We did not expect our decisions to be questioned, nor did this often happen. We made little or no use of external referees, trusting our own judgement even in fields in which we could not really claim to be expert. The simplicity of the arrangements had the great advantage of avoiding

delay and we took pride in being able to offer a speed of publication which I believe compared favourably with that of any other scientific journal of comparable standing. In this we were greatly helped by the speed and efficiency of our publishers, the Cambridge University Press. On the other hand, the lack of any assistance apart from minimal secretarial help did place a considerable burden of routine work on the editors; for example, we read all proofs ourselves, both galley and page, and from this task there could be no let-up during holidays or at any other time, if our reputation for prompt publication were to be maintained.

"Scientifically we undoubtedly took risks in relying so completely on our own judgement, and I am sure that we must have made mistakes. Indeed, I remember two scrapes that I got into myself, one of which caused the resignation from the membership of the Society of a senior continental professor who took exception to an editorial alteration that I had made to one of his papers (fortunately he later returned to the fold); the other occurred when I referred back a paper by a senior biochemist in this country, and as a result had the whole of his department up in arms against me; here again, as it turned out, personal relationships were not permanently impaired.

"Nevertheless, incidents of this kind were warnings of the more serious results that might ensue from editorial misjudgement, and at the same time the likelihood of such misjudgement was rapidly increasing owing to the rising flow of papers for publication and the broadening of the subject matter. For this reason Harden and I persuaded the Committee to allow us to recruit more editorial colleagues. We naturally sought for men who were expert in the fields with which we ourselves were less familiar and we were fortunate in obtaining the help first of all of F. J. W. Roughton to deal with papers involving physics and physical chemistry and later of S. J. Cowell and Frank Dickens to cover the fields of nutrition and of cellular biochemistry respectively.

"With these accessions we were able to carry on reasonably well for a few more years, but there still remained the problems of proof-reading and indexing with which we had no assistance and which were becoming more burdensome with the continuing increase in the flow of material. In 1942 I was appointed Director of the National Institute for Medical Research and had perforce to give up my editorship; this afforded the opportunity for the Committee to consider how they wished the *Journal* to be conducted in the future. The decision was made to appoint an enlarged editorial board, and at the same time to introduce certain changes of policy, among which the most important was the use of external referees to help in the assessment of papers for publication as a matter of routine rather than as a procedure reserved for specially difficult cases.

"These changes were the beginning of the development of the substantial organization that the Society now employs for the production of the *Journal*. The changes were inevitable and were probably overdue. They did, however, come in time to enable the *Journal* to keep pace with the enormous increase in biochemical research that has occurred during the past twenty-five years and to strengthen its position as one of the leading scientific journals of the world. That this should be the outcome is a more than adequate reward to those members of the Society who did their best to maintain the standards of the *Journal* so long as the task remained within the scope of amateurs."

The overall statistics indicating the growth of the *Journal* from 1906 (under the control of Benjamin Moore 1906–1912) until the end of Harington's term of office are given in Fig. 6.1, where the numbers of papers published are recorded. The number of pages published has also increased proportionately but are not recorded because occasional changes in format does not allow direct comparison over the whole period under consideration. With the exception of the period of the Second World War there has been a steady increase in the number of papers published.

During 1945–1965, when the size and print number of the *Journal* increased considerably, the perennial problems associated with publishing an expanding *Journal* arose: the difficulty of costing because of the unpredictable size of each volume, the problem of setting an appropriate level for non-members' subscriptions and the difficulty of obtaining the agreement of the membership to increase their fees. The large profit apparently made by the printer and publisher [the

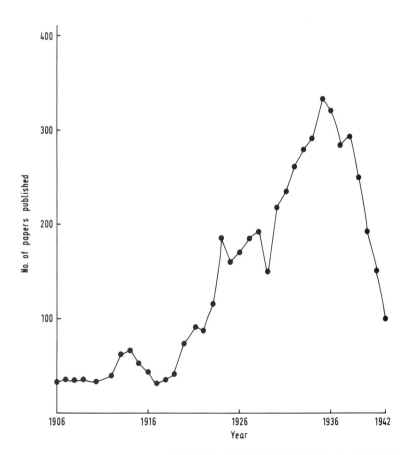

Fig. 6.1. Annual number of papers published in the *Biochemical Journal* from 1906, when it was founded, until 1942, when Sir Charles Harington retired from the post of Editor.

Cambridge University Press (C.U.P.)] did not help matters. All these problems arose again in 1965–1985 but on a much larger scale and are discussed in detail in Chapter 4. However, they were there right from the start. From 1913 to 1920 the overall profit on the *Journal* was £383.18.1 (£383.90) [c. £7000] but it has to be remembered that all the editorial activity over that period was unpaid. The C.U.P. agreed in 1920 that "all profits in the *Journal* is the property of the Society, but is subject to the right of the Press for certain considerations received, to veto what it may consider an improper use of the money"; apparently this represented a concession to the Society [1]. In 1920 The Royal Society donated £50 [£900] for the publication of a series of long and important papers and in 1922 Professor O. Warburg paid for the publication of his papers.

The continual but legitimate demand by the *Biochemical Journal* for additional pages came to a head in 1921 and an appeal was made for funds. It resulted in donations of ten guineas (£10.50) [£280] from Glaxo Ltd and British Glues and Chemicals and of five guineas (£5.25) [£140] from Mr Chaston Chapman. The costs continued inevitably to rise and in 1923 the C.U.P. agreed to a new arrangement which was slightly more favourable to the Society: their commission was set at $12\frac{1}{2}$% on both sales and printing costs and the Society would receive all profits and the right and responsibility to fix the price of the *Journal* and of reprints. Because of this increased responsibility placed on the Society an Editorial Committee was set up, consisting of the Society Chairman, the Honorary Secretary, the Honorary Treasurer and the Editors, to oversee developments. In spite of these changes the extra pages in volume 17 (1924) used up all the Society's profits and by 1925 a further increase in size of 500 pages caused a loss of £33 [£650] that year. A last minute grant-in-aid by The Royal Society saved the day. But the problem would not go away and in 1927 the *Journal* drew on a further £150 [£3000] of the Society's funds. The large commission charged by the C.U.P. was now considered the main cause in this continued financial instability. The Press again made concessions to representations made by a high-powered visiting group consisting of Professor [Sir Rudolph] Peters (Fig. 3.11), J. A. Gardiner (Fig. 2.1), Sir Arthur Harden (Fig. 1.5) and Sir Robert Robinson. The rebate on printing charges was increased from $7\frac{1}{2}$ to 12% and the $12\frac{1}{2}$% commission on members' copies was waived. This change, made retrospective for 1928, saved £243 [£5000], about 10% of the annual cost of producing the *Journal*.

However, the obvious solution, which had been looming for some time, was adopted in 1931 as the result of a projected deficit of £400–£500. The annual subscription for members was increased to 2 guineas (£2.10) [£50] and for non-members to 3 guineas (£3.15) [£75]. This year also saw the first discussions in Committee of a possible move away from the C.U.P.

An important development in 1934 was the decision to publish the *Journal* monthly instead of bimonthly.

So, after some 20 years of what might reasonably be called a "hand to mouth existence" sustained by dedicated Editors, the *Biochemical Journal* emerged as a well established publication and an acceptable if unexciting consolidation period lasted until the end of the Second World War. Indeed the finances were such that the Society did not have to call on funds made available in 1941 to The Royal Society by the Rockefeller Foundation for those societies which were finding difficulty in keeping their learned journals going in war time. In contrast, £1000 [£17,000] was transferred to the reserve fund. When Harington's resignation in 1942 was accepted a special sub-committee [J. H. Bushill (Fig. 5.1), Sir Frank Young (Fig. 3.12), N. W. Pirie, W. T. J. Morgan (Plate 4A), B. C. J. G. Knight and Sir Jack Drummond] was constituted to consider the future of the *Journal*. As a result of their recommendations the main Committee agreed that the affairs of the *Biochemical Journal* should be run by an Editorial Board of about six, with one member being designated Chairman and the remainder Members of the Editorial Board; these replaced the previous Editor and Assistant Editors. This basic arrangement, albeit enlarged, exists to this day. Honoraria were abolished but effective secretarial assistance was provided for the Chairman who dealt directly with authors on behalf of the Board. The Chairman became *ex officio* a member of the main Committee as did one member of the Editorial Board, annually, in rotation. A recommendation, which was to save the Society considerable amounts of money in the future, was made by N. W. Pirie; he proposed that volumes should not expand to cope with the papers available but should be confined to 600 pages. F. G. Young was elected the first Chairman of the Editorial Board in 1942. Six members of the Editorial Board were also appointed; they were chosen so as to cover the major fields of Biochemistry. Today (1986) there are some 50 members of the Board as well as four Deputy Chairmen and an Advisory Panel of over 250 members.

Professor E. J. King (Fig. 3.2) replaced F. G. Young in 1946 and an Honorarium of £200 [£3000] p.a. was introduced for the Chairman. King ran the *Journal* with the scientific help of Dr W. Klyne and Dr I. D. P. Wootton from his Department at the Post-Graduate Medical School at Hammersmith. It was not until 1950 that honoraria of £50 [£600] were voted for members of the Editorial Board. Today Board members are paid *pro rata* for the work they do. It is a complicated arrangement but allows for the size of the Board to be increased without increasing the overall cost.

There had always been times when the Society had been dissatisfied with the C.U.P., either from the point of view of speed of production or cost of production, or because of its inflexible attitude to what appeared to be reasonable requests. Serious problems arose later, as we shall see, but one unexpected confrontation in E. J. King's time was recalled by Dr R. L. M. Synge, who was a member of King's Editorial Board. In Synge's own words:

> "He (King) had been asked by the Committee to find printers cheaper than the C.U.P. and thought he had succeeded, when some smallish printers somewhere in E. of England had tendered at $\frac{2}{3}$ the rate charged by C.U.P. and had set up a creditable sample $\frac{1}{2}$ sheet from TSS rich in figures, symbols and green ink. With the sample in his pocket, Earl King went to Cambridge to see a high managerial personage (HMP) in a sombre room of the Pitt Press, Trumpington Street. He started by saying he thought the C.U.P. were overcharging the Society.
>
> HMP I realize we're pricey, as printers go, but you'd have to look a long way to find someone who'd do as well with all those symbols and formulae in the copy you send us.
>
> EJK (drawing sample from pocket): Well, what do you think of this?
>
> HMP (having examined sample): Do you mind if I take it over to the window for a closer look? (Does so, peering at it through a magnifying glass).
>
> HMP (returns from window, hands back sample): I'm sorry — *we* did that.
>
> Interview ends, neither party having any more to say."

King was followed in 1952 by Professor A. Neuberger (Plate 3A), by which time the number of Editorial Board members had risen to 13. Again the *Journal* office was in rented accommodation in the Chairman's place of work, the National Institute for Medical Research at Mill Hill (see also Chapter 2).

When Neuberger resigned from the chairmanship on moving from Mill Hill the Committee agonized about the possibility of appointing a full-time Editor. They eventually decided to continue with the same arrangements as before but to provide day to day help by appointing a 'full-time Editorial Assistant' with appropriate experience in Biochemistry or Organic Chemistry to deal with routine and technical matters. Early in 1955 Mr F. Clark (Fig. 6.2), was appointed to this post under the title 'Secretary to the Editorial Board', and was in the post when the new Chairman, A. G. Ogston (Plate 2B) took over from Neuberger. Ogston, who acted from 1955 to 1959, travelled from Oxford every Tuesday to spend the day on *Journal* business. He instituted the post of Deputy Chairman. The first holder was Dr T. S. Work, later to be Chairman of the Society's main Committee (Plate 1A). The other members of the Board now numbered 14.

Fig. 6.2. Mr F. Clark. Secretary to the Editorial Board, 1955-1968.

One of the first requests to Ogston from the main Committee was to consider whether the Society might become its own publisher. The Committee was still concerned over its arrangements with the C.U.P. and had noted that The Royal Society had successfully taken over publication of their journals from the publishing house. Together with Work and Clark, Ogston discussed the *Journal's* problems with Officers of The Royal Society and it was agreed that a similar arrangement for the Society would be profitable in the long run, if not immediately, although considerable administrative reorganization would be necessary. The main Committee's reaction was, however, cautious and no moves had been made when in 1956 Pergamon Press wrote offering to open discussions about publishing the *Journal* more cheaply and efficiently than the C.U.P. Investigations into this possibility were undertaken but it was decided not to take up the offer.

In a recent letter Dr Ogston has pointed out that, during his chairmanship:

> "the Editorial Board was collectively a happy and harmonious body; individually, I was impressed by the care and attention that Editors gave to the interests of authors as well as to those of the *Journal*. Characteristic of this (although an extreme example) was the action of one member of the Board who, over many months, corresponded with and visited an author, making suggestions for confirmatory work which resulted in great improvement of the paper".

The efforts of the Editorial Board at that time mirrored the attitude of its Chairman, who was so concerned about unnecessary misunderstandings with authors that he spent much time trying to devise ways to deal with this. An ingenious solution suggested itself:

> "Much of the (never very serious) dissatisfaction that authors felt about the Editorial Board arose, I believe, from their ignorance of the editorial process and its aims, and I never lost a chance to do what I could to explain them. This led me to the idea that we might make a facsimile booklet to demonstrate this process, editors' reports, Chairman's letters and all, but it was too difficult. Our specimen paper would have (if it were to spill the whole beans) to be acceptable, but to require extensive revision. I could not imagine the author of any 'real' paper of this kind being willing to have it publicly exhibited in this way. So a phoney paper would have to be used, inserted (without Editors knowing it) at the start of the process and withdraw from it before going to the Press. I found I could not devise such a paper."

Three main issues which exercised the Board during Ogston's term of office (and presumably to varying degree at most other times in the Board's existence) were: "how to keep costs down and how to keep down the interval between submission and publication and what should constitute 'Biochemistry' in relation to the subject matter of papers being judged suitable for the *Journal*". The result was an agreed statement that the *Biochemical Journal* "should publish papers in all fields of Biochemistry — plant, animal and microbiological — provided that the results make a new contribution to biochemical knowledge; or that they describe methods applicable to biochemical problems". In spite of this public assertion of policy many plant biochemists felt, rightly or wrongly, that they were discriminated against. In the early 1960s the Plant Phenolics Group widened its horizons and became the Phytochemical Society and founded the journal *Phytochemistry* with the help of a commercial publisher. This developed into a flourishing international publication, which celebrates its 21st birthday in the same year as the Biochemical Society celebrates its 75th anniversary. If the Biochemical Society had been a little more adventurous in the early 1960s it might have had another prestigious journal under its wing. The nagging feeling that plant Biochemistry has not been well treated certainly persists to this day, although, to insert a personal note, I have never found it so. But there is one eminent member of the Society who would not agree with this and who has not published in the *Biochemical Journal* for many years.

In 1959 Ogston was succeeded by Professor W. V. Thorpe (Plate 2B), whose period of office until 1963 was difficult mainly because things came to a head with C.U.P. The *Journal* was expanding rapidly but a new financial arrangement, proposed in 1961 by the Press, was in no way to the liking of the then Treasurer (F. A. Robinson; Fig. 4.2). The situation was described by R. A. Morton [1]:

> "The Treasurer had reported early in 1961 that the Cambridge University Press proposed a new financial arrangement to be operated from 1 January that year. The commission of 15 per cent on all sales of the *Journal* and other publications would remain as previously. The commission on cost of production of the *Journal* had been $2\frac{1}{2}$ per cent but the proposed new basis was 'a commission of $3\frac{1}{2}$ per cent of the volume price (or where there was no volume price the aggregate prices of the parts) multiplied by the number of copies of the volume being printed'. It was calculated that on the figures for 1959 the Society would have saved about £900. The Treasurer was instructed to look into the effect of the new proposals and, after analysing the figures for 1958, 1959 and 1960, and extrapolating to 1962, he reported that the financial trend of the proposals was unfavourable to the Society. He would have preferred a sliding scale based on the old system whereby the percentage commission could be progressively reduced as the circulation increased."

With an average print run of 7000 copies the Society felt that it should have been given better terms but the Press was adamant; they believed that the successful journals they published should subsidize the less successful ones. Apart from this the loss of about £1500 [£11,000] made by the *Journal* in 1961 was, according to W. J. Whelan (Fig. 3.10), then Honorary Secretary, due to the absurdly low rate set by C.U.P. for the cost of separates: "Some three or four years later, when we had left C.U.P. ... the true cost was found to be ten times that C.U.P. were charging".

If the Society were to become its own publisher there was clearly a difficult time ahead. However, a life-line appeared in March 1962 when the Chemical Society, now the Royal Society of Chemistry, which had recently set up a distribution centre, offered to distribute the *Biochemical Journal* for £2700 [£20,000] per annum and to store back numbers for £250 [£1800] p.a.; the corresponding figures for the C.U.P. were £8000 [£58,000] and £400 [£2900], respectively. This convinced the Committee that considerable economies, and possibly profit, would result if the Society became its own publisher. However, the C.U.P. was not willing to print the *Biochemical Journal* if it were not also the publisher. On the other hand they agreed to continue with the existing arrangements until new printers could be found. By June 1962 the Committee had decided to break with the C.U.P. and an active search was made to find an appropriate publisher so that the new arrangements could begin in January 1964. The Editorial Board were most unhappy over these developments, for they felt that "they were being treated more as junior employees than equal partners whilst they were, at the same time, aware that the sales of the *Journal* underpinned financially the expanding activities of the Society" [2].

The proposed change which would involve the loss of the great experience of the C.U.P. "reader" for detailed editing would, they felt, inevitably result not only in lowering the high standards set by the Editors but also in causing them a great deal of extra work. These points, unjustified in the event, and others were put to the Committee in November 1962 by Thorpe, who felt considerable personal loyalty to the C.U.P. The financial advantage which, according to the protagonists of change, would accrue from employing the proposed new printers was also challenged. After a long and heated debate the proposal to leave the C.U.P. and to employ new printers was carried by eight votes to six. After the voting the Chairman, Professor J. N. Davidson (Fig. 3.14), indicated that he strongly supported the proposal. At this meeting Thorpe's imminent retirement, after ten years of devoted service to the *Journal*, was reported.

The Editorial Board met shortly after the November meeting of the main Committee; they "read with interest statements by the officers in favour of the change of printers but remained unconvinced about the wisdom of the change". As they considered themselves no longer sufficiently independent to conduct the business of the *Biochemical Journal* the Board decided to resign *en bloc* from 1 January 1963. They ameliorated this uncompromising position somewhat by agreeing to continue in an acting capacity until a new Board could be constituted. This interim arrangement was to be as short as possible and would not extend beyond 1 September 1963.

At the Committee meeting on 11 December Dr H. J. Rogers (Plate 3B), the Deputy Chairman of the Editorial Board, who was shortly to be confirmed as Chairman, presented the Board's case after which Davidson emphasized that the Rules of the Society clearly indicated that the ultimate responsibility for the management of the Society's affairs lay with the elected members of the Committee. The point was clearly made that the Editorial Board was under the general jurisdiction of the main Committee. After much emotional debate it was agreed that four members of the Committee and four of the Editorial Board should meet as a working party under the chairmanship of Professor N. F. Maclagan on 14 December to seek a way of dealing with the impasse. The proposed compromise to defer the arrangements for one year satisfied the Board members and was accepted by the Committee members with, one suspects, some relief. The printers with whom they had made preliminary arrangements turned out not to be big enough for the job. The working party also recommended that an Advisory Committee for Publications be set up as a co-ordinating body.

The Editorial Board accepted the proposals of the working party and the Advisory Committee for Publications (ACP) was set up; its constitution is given in Chapter 3. At its meeting on 20 September 1963 the Committee considered the unanimous recommendation of the ACP that as from 1 January 1964 the *Biochemical Journal* be printed by Wm Clowes (later Spottiswoode, Ballantyne Co. Ltd) and published by the Society using the Chemical Society as its agents. It was calculated that this would result in savings of some £3600 [£24,000] in 1965. The Editorial Board did not object to this arrangement and the Committee put it into action with all speed. The Chairman of the Board (Dr Rogers) played a big role in bringing these discussions to a satisfactory conclusion. Thus ended one of the most difficult problems the main Committee has ever had to face, but there is no doubt that the final outcome was advantageous to the Society. It is ironic to find that only very recently (October 1985) the printing of the *Journal*, volume 231, has reverted to the C.U.P., who some 22 years on do an excellent printing job economically but now with no publishing strings attached.

6.3 The *Biochemical Journal* 1965–1986

The trauma of the changes in publishing the *Biochemical Journal* demanded a period of quiet consolidation and this was provided under the chairmanship of Dr H. J. Rogers. However,

Fig. 6.3. Dr W. N. Aldridge, O.B.E. Chairman of the Editorial Board of the *Biochemical Journal*, 1965-1969.

his successor, Dr W. N. Aldridge (1965-1969, Fig. 6.3), realized that further pressure was building up on the Editorial Board with the rapidly increasing number of papers being submitted in ever-widening aspects of Biochemistry. He agreed to take on the job only if the number of Editorial Board members were doubled, from 18 to 36. He wrote: "You will be amused that this was done so rapidly that at my first Board meeting we had to wear name tags". During this time, the number of Deputy Chairmen was increased from one to three and these were the nucleus of the Editorial Committee established by Aldridge. This met more frequently than the Editorial Board and enabled detailed technical decisions to be made quickly (the Editorial Board meets only twice a year), so that publication time could be reduced to a minimum. The relatively long publication time was considered one of the main reasons for the *Biochemical Journal* then not attracting papers in the area of so-called Molecular Biology, which at that time was alive with exciting observations. Delayed publication time was certainly one reason but it was the enthusiasm and drive of the young molecular biologists who wanted to get their results to the widest audience of like-minded specialists which led them to eschew general, archival types of journal and to form new specialized journals; at that time the *Journal of Molecular Biology* was a favourite outlet. The problem remains up to the time of writing and one notes that many 'molecular biology' investigations are first reported at Society meetings but the ensuing substantive papers do not appear in the *Biochemical Journal*.

Professor D. G. Walker (Plate 2B), Aldridge's successor, continued the drive to cut down publication time and succeeded in attracting so many papers that the increasing size of the *Journal* sometimes caused financial tremors in the Society Committee (see Chapter 4). During this time the ACP was briefed to search for possible new Society publications. A detailed proposal for a new "Journal of Sub-cellular Biochemistry" was considered but the perceived possible overlap and competition with the *Biochemical Journal* resulted in its rejection. However, an important compromise emerged: the *Journal* was sectionalized so that alternate issues were devoted to *Molecular Aspects* (blue cover) and *Cellular Aspects* (orange cover) respectively. This not only emphasized the widespread coverage of the *Journal* but allowed members to subscribe to one half of the *Journal* at the run-on cost. This was an important concession as printing costs and thus subscription rates were rapidly increasing. The sectionalization continued for 11 years and only recently (1985) have the two parts been re-combined; however, a sectionalized contents page has been retained.

During Professor Walker's period of office the *Biochemical Journal* lost a faithful servant when the Editorial Secretary, Mr

Frank Clark (Fig. 6.2), was killed in a road accident in 1968. He dealt with all aspects of the day to day activities of the Editorial Office with great efficiency and dedication; he was very involved with the transfer of the *Journal* from the C.U.P. to new printers. Frank Clark was succeeded by Dr J. D. Killip.

Two developments during Dr J. Dingle's (Fig. 6.4) period as Chairman of the Editorial Board (1975–1982) were of particular importance. One was the reorganization of the Editorial Office, which had to be carried through under a cloud of staff problems. In 1978 Mr A. (Tony) G. J. Evans (Plate 2A) was appointed Editorial Manager and later Dr A. S. Beedle was recruited as Deputy Editorial Manager with special responsibility for the *Biochemical Journal*. These appointments and the resulting new procedures in journal management combined to produce a more effective editorial unit, which remains in being at the time of writing.

Fig. 6.4. Dr J. Dingle. Chairman of the Editorial Board of the *Biochemical Journal*, 1975–1982.

The second development stemmed from a suggestion from the Committee that handling charges should be instituted as a way of dealing with financial problems. This idea was entirely against the publication ethos of British science in general, based as it is on the right of free publication subject to peer review. It is difficult to decide whether the Committee proposal was a serious suggestion or coat trailing. The Editorial Board not unexpectedly rejected the, to them, preposterous idea out of hand but the Chairman did set up a small sub-committee to look into the procedures for handling papers and the opportunities for further streamlining editorial activities. The outcome was a number of far-reaching proposals which have proved highly beneficial; they include (1) the introduction of a panel of 250–300 expert Editorial Advisers who are given free membership of the Society in return for agreeing to review up to ten papers a year (about one-third of the advisers are from overseas, thus helping to emphasize the international image of the *Journal*); (2) a speeding up of reviewing so that decisions on papers are given within 6–8 weeks of their receipt in the Editorial Office; (3) the introduction of Reviews and *B.J.* Letters, of which more later; (4) the agreement that the Editorial Board should be internationalized (currently 11 of the 50 Board members are from overseas). The continual fight to reduce the publication time has, with occasional hiccoughs, over the past 30 years been successful (Fig. 6.5): the delay in the 1950s was some eight months; in the 1980s it is just a little more than six months.

Recent recommendations have speeded up the aim to project the *Biochemical Journal* as an International Journal of Biochemistry. This began with the institution in the 1970s of overseas advisers, who have now been subsumed within the Editorial Board. This development has resulted in the "love–hate" relationship, as one recent member of the Board put it, between the main Committee and the Editorial Board,

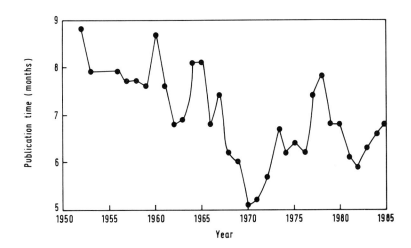

Fig. 6.5. Variation in average publication time of papers submitted to the *Biochemical Journal* between 1950 and 1985.

Fig. 6.6. Professor C. I. Pogson. Chairman of the Editorial Board of the *Biochemical Journal*, 1982–1987.

the former maintaining that the production of the *Biochemical Journal* is only one aspect of the activities designed to give maximum service to Society members, whereas the latter, in general, maintains that as a well established international journal it need not be closely associated with a national society. During Professor Pogson's (Fig. 6.6) chairmanship (1982–1987) the *Journal* had a stand under its own banner rather than that of the Society at the FASEB (Federation of American Societies for Experimental Biology) meeting in 1984 at St Louis and in 1985 at Anaheim (CA) because of "the difficulty of trying to project an international image for the *Journal* with a national image for the Society". The answers to a questionnaire distributed at the FASEB meeting and at the IUB meeting at Amsterdam (1985) have been processed and the conclusions reported in the *Biochemical Society Bulletin* [3].

This schizophrenia will undoubtedly continue into the future but one is left wondering what deleterious effect, if any, the sponsorship by the Biochemical Society has on the already impressive international image of the *Biochemical Journal*. Arguments about umbilical cords aside, there is no doubt that the *Journal* is attracting many more overseas contributions (Table 6.1). Considerable effort has been made in the last few years to dissipate the belief which has arisen over the years that the somewhat rigid attitude of the Editors to relatively minor problems of presentation has discouraged authors from submitting their more exciting papers to the *Biochemical Journal*. Authors are now allowed more stylistic freedom and an extended general use of abbreviations, and they can now choose to give their references either in the Harvard system in which references given as, for example, Jones & Smith (1984) in the text are listed alphabetically at the end of the text, or in the numerical system, that is sequential numbering in the text corresponding to the numbered list of references. Running

Table 6.1. International origins of papers published in the *Biochemical Journal*

Year	U.K. (%)	U.S.A. (%)	Europe (%)	Elsewhere (%)
1920	73.3	1.3	9.1	15.3
1930	80.5	3.6	3.2	12.7
1940	81.1	6.3	3.2	9.4
1950	88.6	0.9	1.4	9.3
1960	69.4	5.9	4.4	20.8
1970	60.9	13.9	7.3	14.9
1980	45.2	22.1	15.9	16.7
1986	41.7	29.4	22.1	6.8

parallel with this increased freedom of style is the requirement for authors to write as succinctly as possible, a requirement enforced by the current practice that, except under special circumstances, papers should be no longer than eight printed pages. This restriction also allows the Board to keep within the size limits laid down by the Committee without refusing good papers. The recent return to the C.U.P. as printers has been accompanied by many stylistic changes to give the *Journal* a modern look. The 'desectionalization' of the *Journal* in 1984 was achieved only after prolonged discussion, not least over what colour the cover of the merged *Journal* should be. An important result of the merger is that as the *Journal* now appears every fortnight any paper just missing inclusion in an issue is held back only two weeks. Previously the delay would have been one month, because each section appeared only monthly. This represents one victory in the constant battle to improve publication time. Apart from these efforts more conventional public relations ploys have been used in attempts to increase the sales of the *Journal*, particularly in the U.S.A. and Japan. To this end Dr G. A. Snow (Fig. 6.7) was appointed Promotions Organizer in 1976. He and his colleagues mounted a large exercise to attract new subscribers. Personal letters were sent to some 60 named individuals in various institutions and resulted in three new subscriptions; this was regarded as a good yield but "it was time-consuming and could not easily be repeated". Snow went on:

"A library will subscribe to the Journal if there is sufficient insistence from the practising scientists within the institution. The librarian is always faced with conflicting demands which have to be met from a limited budget, and will respond according to the urgency of the demand by the users. There will always be places where interest is marginal and subscriptions are liable to be cancelled when funds are scarce. To some extent those losses are balanced by unexpected new subscriptions from places where there has been an upsurge in biochemical activity.

"To a large degree the *Journal* sells itself on its scientific reputation. It cannot be treated as a commercial commodity and sold by

Fig. 6.7 Dr G. A. Snow. First Promotions Organizer, 1976.

skilful persuasion. Advertising has little place in promoting the *Journal*. One intractable problem is to whom promotional material should be addressed. Librarians receive shoals of leaflets and brochures. At best they put them on display for a time; most go directly into the waste paper basket. Directors of institutions rarely have any direct interest in Biochemistry, and will treat advertising material with indifference. Working biochemists will already know of the *Journal* and need no reminder."

In relation to the question of promotion, it has recently been agreed that it is not worthwhile for the Society's publications to be separately represented at the meetings of the American Library Association. Attempts to "promote" Molecular Biology papers have been made by distributing a leaflet to members of the Nucleic Acid and Molecular Biology Group and members of EMBO. The assessment of the results is not yet available.

6.3.1 Rapid Publications

Accelerated publications were first introduced in 1964 in order to attract significant new work to the *Biochemical Journal*. They were named "Short Communications" and were printed at the end of each issue of the *Journal*. The era of rapid publication journals was emerging and, in spite of the enthusiasm of the Advisory Committee for Publications for launching a "quickie journal", the conservatism of the Editorial Board and General Committee carried the day and the idea that the Society should publish such a journal was not accepted. It is not often during its history that the Society has missed a clear opportunity of being one of the first in the field. Eventually this gap was filled in Europe by the launching in 1968 of *FEBS Letters*, which has been a great success, has maintained high standards, thanks to Professor S. P. Datta, a member of the Society, who was Editor from 1968 to 1985. It has had no obvious ill effects on the FEBS archival journal, *European Journal of Biochemistry* (*EJB*). It is interesting to note that the Editors of *EJB* opposed the publication of *FEBS Letters*, but presumably the FEBS Publication Committee had at that time more teeth than its Biochemical Society counterpart.

The Society's compromise reaction to the move towards a rapid publication journal was, as just indicated, the inclusion of Short Communications within normal issues of the *Journal*. In 1968 the Communications were revamped into the form in which they exist today — "Rapid Papers". The publication time of Rapid Papers is about one half that of normal papers (e.g. 16.3 weeks and 27.8 weeks respectively in 1983) and they now represent some 12–13% of the total papers published, whereas in the mid-70s the percentage was around 17–18%. The only difference between Rapid Papers and full papers is the length — the former must not occupy more than four

Journal pages. The quality of the publication is the main criterion for acceptance — "urgency" is not one, being a very subjective assessment. In fact any submitted paper of four or fewer pages is treated as a Rapid Paper with the authors' agreement.

6.3.2 *BJ* Reviews and *BJ* Letters

In 1980 it was decided to introduce these two new types of paper into the *Biochemical Journal*, thus reversing a categorical rejection of Reviews by the Editorial Board in 1964. The Editorial Board was fortunate to persuade Professor J. A. Lucy (Fig. 6.8), who had just retired as a Deputy Chairman of the Board, to accept the job of pioneering this venture. He writes:

Fig. 6.8 Professor J. A. Lucy. Deputy Chairman of the Editorial Board of the *Biochemical Journal*, 1979–1986.

"Initially, there was some apprehension among individual readers and contributors to the *Biochemical Journal* that the publication of review articles would increase still further the existing pressure on space in the *Journal*. It was, however, not intended by the Editorial Board that reviews should occupy more than a very small part of the *Journal* and, in the event, the review articles have proved to be popular with research workers, university teachers and students. Indeed, one student at the University of Surrey was able to quote extensively in an oral examination, for the benefit of the external examiner, from a review only some three weeks after it had appeared in print! Initially, also, rather pessimistic forecasts were made that, because of the number of review journals now being published, it would not be possible to attract good reviews to the *Biochemical Journal*. Fortunately, this has never been the case, and reviews are in fact now being published more frequently than they were at the outset. Although some difficulty was experienced at first in commissioning reviews because prospective authors occasionally feared that their articles might not be as widely read as they would like, this ceased to be a problem after about two years, and approximately one third of the reviews now being published are actually suggested by prospective authors. A majority of the reviews are nevertheless still commissioned.

"*BJ* Letters provide an opportunity to discuss, criticize or expand particular points made in published work, or to present a new hypothesis. At the time that *BJ* Letters were initiated, the Editorial Board decided that — when a Letter is polemical in nature — a reply may be solicited from other interested parties before its publication. This has proved to be an interesting feature of the Letters, and a number have been published simultaneously with a reply from an interested party. Ding-dong counter replies, and counter–counter replies, of the kind that feature in some other publications are, however, not published in the *Journal*. Although tact is required in handling the occasional, abrasive communication, a majority of the submissions received are written in the spirit of discussion that the Editorial Board wished to encourage as a feature of *BJ* Letters, and the Letters appear to be fulfilling a useful function, since the number of submissions is increasing."

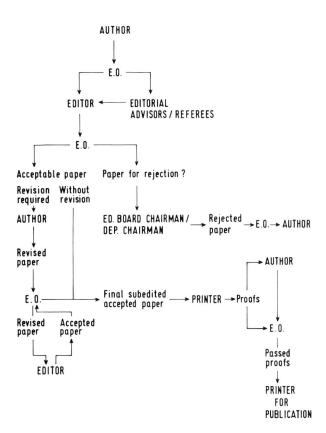

Fig. 6.9. Flow sheet for the passage of a paper submitted to the *Biochemical Journal* through to its final preparation for publication. E.O., Editorial Office.

6.3.3 Editorial Office

In keeping with the highly professional production of the *Biochemical Journal* the Editorial Office is a very efficient organization and the flow-sheet (Fig. 6.9) indicates how submitted papers are dealt with.

In spite of all the problems since 1945 just described the *Biochemical Journal* has sailed serenely on with an ever-increasing number of papers being submitted and published (Fig. 6.10); submissions of Rapid Papers is also increasing but the number accepted for publication is now steady at around 100 per annum (Fig. 6.11); the Reviews have established themselves as authoritative expositions of critically important areas of Biochemistry. All this adds up to a prestige *Journal* which continues to provide substantial income for the Society.

6.4 *Clinical Science*

On 10 and 11 April 1954 the Association of Clinical Biochemists and the Biochemical Society held meetings in Edinburgh on successive days and gave publicity to each other's meetings so that members of either Society could attend both meetings. From this arose the idea discussed

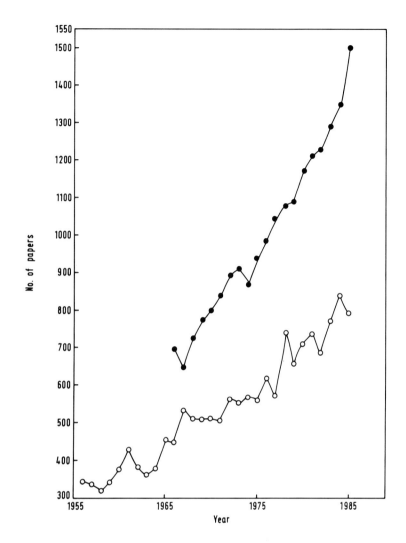

Fig. 6.10. Number of full papers submitted (●——●) to the *Biochemical Journal* and number accepted (○——○) over the period 1955 to 1985.

formally in 1956 that a 'Journal of Clinical Chemistry' be established and in 1957 the Society responded positively to a memorandum recommending this circulated by the then Honorary Secretary (Dr C. E. Dalgliesh; Plate 4C). Meanwhile, the Medical Research Society had approached the Biochemical Society with the suggestion that its journal, *Clinical Science* (which was founded as long ago as 1909 under the title *Heart*), should be broadened with the Society's collaboration. The Association of Clinical Biochemists agreed to widen the discussions with the Society to include this new proposal and in October 1957 the following proposals were recommended:

(i) that *Clinical Science* should continue to be the medium for the publication of papers primarily on diseases of man

(ii) that papers on pure methodology would not in general be accepted

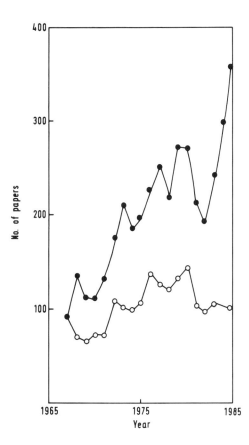

Fig. 6.11. Number of 'rapid' papers submitted (●——●) to the *Biochemical Journal* and number accepted (○——○) over the period 1967 to 1985.

(iii) that there should be parity of editorship between the Medical Research Society and the Biochemical Society

(iv) that the Medical Research Society should recommend the Trustees of *Clinical Science* to increase their number to four, two of whom should be representatives of the Biochemical Society

(v) the Trustees would be the legal owners but would have no concern with the day to day running of the journal.

These proposals were generally accepted by the Committee and the necessary legal agreements, which basically meant that *Clinical Science* would be run jointly and the profits would be shared equally between the two Societies, after deducting charges for work of the Editorial Office and administrative overheads, was ratified in 1960. Four Trustees, two from each Society, were appointed and a Committee of Management set up which consisted of the Honorary Secretary (the Senior Secretary if more than one) and the Honorary Treasurer of each Society together with Chairman and Deputy Chairman of the Editorial Board and one other representative from each Society. The Editorial Board was made up of four persons from each Society with the possibility that one of the Biochemical Society's members could be nominated by the Association of Clinical Biochemists. The Societies were to agree on the appointment of the Chairman and Deputy Chairman so that they did not both represent the same Society. The maximum term of service on the Board was to be five years which could be increased to eight if the member were Chairman or Deputy Chairman at the end of his five year stint. There was also a clear statement in the Agreement that in the case of any conflict between the Editorial Board and the Committee of Management the view of the Committee of Management would prevail. The Trustees at the present time are the Biochemical Society itself, as a limited liability company, and two individuals, Sir John McMichael and Sir Melville Arnott, appointed by the Medical Research Society.

Each Society contributed £1000 [£7500] to a joint account to finance the journal and arrangements were made with Messrs Shaw & Sons to print it and with the C.U.P. to publish it. Publication under a joint Committee of Management eventually began in 1962. The problems the *Biochemical Journal* had with the C.U.P. in the 1960s was reflected in the administration of *Clinical Science* and in 1963 the Society Committee accepted a recommendation from the Committee of Management of *Clinical Science* that as from 1 January 1965 the publishing should be undertaken by Blackwells Scientific Publications Ltd.

In 1965 the membership of the Committee of Management was enlarged by increasing the non-office bearing representatives from each Society from one to two. In 1977 a slight amendment was that in the case of the Biochemical Society, the Secretary should be redefined as the Honorary Publications Secretary (the Chairman of the Publications Board).

As indicated above the Editorial Board began with eight members, four representing each Society, but, with the ever-increasing number of papers to process, is now 35 (the maximum agreed with the Committee of Management). In December 1977 the Committee accepted the reality of the difficulty of maintaining an exact numerical balance in the composition of the Editorial Board and agreed that this parity need not be strictly observed provided a "reasonable equilibrium was maintained". The Committee also agreed that the Chairman of the Editorial Board could seek new editors from outside the two Societies provided that such editors would accept membership of one of the Societies on joining the Board. Furthermore it was agreed that all appointments to the Editorial Board needed to be ratified by the Committee of Management. Because of the increased work load a second Deputy Chairman was appointed in 1985. All the editorial activities are today carried out in the Editorial Office of the Biochemical Society; it represents about 20% of the work-load of the Office.

A chronic problem which has recently been solved is the commitment to publish abstracts of communications read at Medical Research Society meetings. This took up considerable space (about 70 pages out of an annual total of 1536 allowed by the Committee of Management) which the Editorial Board increasingly felt could be better used for original, refereed papers. The Abstracts are now issued in the form of separately bound Supplements to *Clinical Science* circulated with the journal at no extra cost to subscribers. The rejection rate of submitted papers runs at around 55–60%. This rather high figure does not apparently deter authors from submitting papers because the annual number of submissions is still on the increase (372 in 1980 compared with 488 in 1985). About 50% of the published papers come from the U.K., with around 27% from the U.S.A., 13.5% from Continental Europe and 9% from Australasia. This distribution compares favourably with that of the *Biochemical Journal*, although there has been no overt attempt to internationalize the Editorial Board.

During the first years of the amalgamation *Clinical Science* was losing money, on average about £5000 p.a., but this situation gradually improved until in 1980 it was making a small profit. Around this time the Medical Research Society were seriously questioning the profitability of the journal, particularly since the International Society of Hypertension (which had regularly used *Clinical Science* Supplements to publish their annual meetings Communications) had decided to transfer to a new journal which was launched by a commercial publisher who guaranteed that Society an annual income of £20,000. The officers of the Biochemical Society looked into the matter and decided that with the fullest possible use of newly available printing technologies it would be possible to achieve similar profitability with *Clinical Science*. The newly appointed Chairman of the Committee of Management, Dr D. C. Watts (Plate 3B) accepted this view and advocated it so enthusiastically that a five year contract was agreed. The change in profitability was quick and dramatic, the annual surplus for 1983 was £48,747 and this increased to £56,918 in 1984 and to £59,518 in 1985. At the moment of writing both Societies seem well satisfied with the situation. The numbers of subscribers has also followed the general pattern noted for other journals increasing well until the late 1970s, when the downfall averages some 4% per annum, a characteristic of most scientific journals. However, in 1986 the downward trend was reversed and the number of subscribers showed a small but significant increase.

6.5 *Biochemical Society Transactions*

The length of the gestation period leading to the birth of *Biochemical Society Transactions* (*BST*) well illustrates how the

Society's present administrative structure can lead to delayed action. This is not to argue that in this case time was not required to study all aspects of the problem but to show how things can be held up. It also emphasizes that, contrary to general myth, the honorary officers of the Society cannot easily force through their own views, however much they felt them necessary for the good of the Society, by presenting Committees with *faits accomplis* drawn up after all night sessions "oiled by carafes of red wine".

Although the proposal to move the unedited Proceedings of meetings from the *Biochemical Journal* was defeated at a General Meeting in 1967 the problem still worried the Committee. This enhanced the growing belief that a publication complementing the *Biochemical Journal* was needed to cover satisfactorily the expanding activities of the Society and the increasing developments in Biochemistry itself as well as to provide members with the service they deserved.

In July 1969 the Committee asked the Advisory Committee for Publications (ACP) to consider a feasibility study on a new 'Transactions Like' journal, in spite of reservations made by the members representing the Editorial Board. The feasibility sub-committee set up by the ACP made detailed recommendations based on a paper by Dr D. C. Watts. These proposals were accepted and presented to the Committee with a strong recommendation for action by the then Chairman of the ACP in November 1969. The main proposals were that *Biochemical Society Transactions* should be sold with the *Biochemical Journal* but be free to members of the Society; it would contain expanded reports on Society and Group colloquia, free communications (unedited) to Society meetings, short (edited) communications, as well as special lectures. It was also suggested that the length of the communications be increased to 600 words and that they be reported in Agenda Papers only as 60 word abstracts. Again, a new worry which arose was the possibility of 'double publication'. However, the most powerful objections by the Editorial Board were that edited 'short communications' should occur together with unedited free communications and that the former should be moved from the *Biochemical Journal*. Technical problems as to who should do the scientific editing were also raised. Inevitably the matter was referred back once again to the ACP. Following further discussions the ACP were told by the Committee in July 1970 to press on with the arrangements for publication of *Biochemical Society Transactions*, but in October 1970 it was agreed to put the proposal to an A.G.M. There things rested until the 1971 A.G.M., when the idea was accepted but with the suggestion that implementation should be delayed for a year. By March 1972 a Managing Editor for *BST*, Professor R. B. Beechey (Fig. 6.12), was appointed and after much labour and frustration he launched the first issue in

Fig. 6.12. Professor R. B. Beechey. First Managing Editor of *Biochemical Society Transactions*, 1972–1976.

April 1973. The journal was well received and the decision to distribute it free to members was an excellent idea from many points of view, not least for the wide publicity that ensued (see Chapter 4). However, it soon became clear that it was not financially feasible to continue to distribute *BST* free to members and from 1974 members received the journal only if they subscribed to it. The total number of subscriptions started at around 2200 and this was maintained until 1980, when with the general recession the numbers began to fall (see Chapter 4). As with the *Biochemical Journal*, the subscription group which maintained its numbers were the U.K. subscribers; the biggest drop on the other hand were U.K. members, down some 64%. The percentage drop in overseas member subscriptions over the same period was, in contrast, 47%. Perhaps this reflects once more the relative support given to science in the U.K. compared with elsewhere.

In 1977 Dr D. C. Watts took over the Managing Editorship of *BST* and at the time of writing remains in charge. He provided some personal comments on this period and they give an authentic flavour of problems encountered and overcome. He writes (a slightly shortened version of his original manuscript):

"Having retired from the Editorial Board of the *Biochemical Journal* in 1974 it came as both a surprise and a challenge to take over as Managing Editor of *BST* in 1977. Brian Beechey had the journal in good shape with an overwhelming amount of copy and reviews commissioned into the foreseeable future. All I had to do was sit back and let it all happen! The honeymoon was short-lived. Within the year the news broke that *BJ*, *BST* and the Society were all drifting into the red and I found myself on the receiving end of a string of letters from Rex Dawson [Honorary Publications Secretary, 1973–1980, Plate 4B] demanding financial economies in publication costs. Nothing could be done about existing copy and proposals took the general form of "going quarterly" and cutting the communications back to the old 400 words. I opposed both of these proposals as sounding the death knell for *BST* and bought time by going back to the drawing board to examine every aspect of production. This resulted in the new format, something I had long desired, and a new rigorous code of conditions for submitting communications with charges for more than one Figure or Table and the minimization of proof corrections, which imposed a substantial bill from the printers. At the same time we were able to go from letterpress to offset printing which, with a change in the paper to that used by the *BJ*, enabled us to include halftones in the text and small improvements in presentation such as having a picture of the Special Lecturer at the beginning of his account. At this time also the first hint of the world recession became apparent; nevertheless *BST* remained in the black and made a modest profit.

"A major problem in managing *BST* is trying to keep the page number approximately constant and to the estimate. I have no control over the major component, communications, but can restrict or encourage colloquium reports and commission or omit review material. There is no doubt the Colloquia and special lectures sell *BST* and I have always felt that for too long it has been necessary to restrict the wrong part of the contents. It was a pleasure to encourage more and longer colloquia when the communications showed a substantial decrease in number following an increase in the annual subscription. The last year (1985–1986), however, has seen a number of factors that have resulted in meetings being larger than ever before. The publishing boom may subside, otherwise I may yet find *BST* under fire again to cut production costs in some way (publication of each communication costs about £50, but the right to publish almost unlimited communications seems to have become part of Society tradition).

"Part of my reorganization was to establish a regular publication schedule. This enabled *BST* to be included in *Current Contents*. The financial squeeze on *Current Contents* resulted in *BST* being discarded with the offer that we could be included in the new *Transactions Contents* — an unattractive publication that covers a wide diversity of topics. A long exchange of letters finally resulted in our being readmitted to *CC* but at the expense of modifying the contents list to exclude communications. All communications are abstracted by *Chemical Abstracts* and so should be thrown up by a computer search that uses *Chemical Abstracts* as a literature base. I subsequently discovered that many journals modified their contents pages to comply with the requirements of *Current Contents*.

"Members rightly demand the shortest possible publication time but this does cause problems in relation to the organization of Society meetings. Recent changes introduced by Roy Burdon [Honorary Meetings Secretary 1981–1985, Plate 4B] have changed both the number and timing of meetings, which, in turn, have altered the magnitude of the attendance over and above those outlined above. I now anticipate two large meetings per annum, Christmas and the A.G.M. Time-tabling the publication of these to span two issues each of *BST* has enabled me to hold to my publication schedule and minimize the delay between meeting and publication — by next year a minimum of 5 and a maximum of 7.5 months. During Roy's experimental period, coupled with the publishing boom, publication delay went up to nearly a year. *BST*, contrary to what some members think, has never been a quick publication journal; the inclusion of proofs for authors has prevented that. (We have discussed eliminating proofs on many occasions but it always emerges that scientific accuracy would suffer considerably if we did so, and with now only a small financial saving.) I make these points to indicate the intricate interrelationship between Society organization and running *BST*."

6.6 *Bioscience Reports*

The failure to start a rapid publication journal in the late 'sixties still worried the Publications Board well into the 'eighties, as did the possibility of producing a journal more directly concerned with molecular biology than is the *Biochemical Journal*. In fact a proposal to start a journal 'Cell Genetics' was pursued as far as collecting the names for a possible editorial board. However, the idea was vetoed and there eventually emerged, rather suddenly, what was probably a compromise idea ('Gene Expression' was also suggested) — *Bioscience Reports*. At the main committee meeting on 29 November 1979 the recommendation of the Publications and the Finance Boards that a Rapid Communication Journal be launched was approved. £75,000 [£110,000] was set aside to start the venture and Professor C. A. Pasternak was asked to become the Managing Editor and provide a feasibility study. The completion of the feasibility study was reported at the Committee and the go-ahead was given for a start in January 1981 within the recommended budget (the feasibility study proposed a higher figure). The journal was to be called *Bioscience Reports* and would print communications and reports in molecular and cellular biology. A prestigious international Editorial Board was quickly appointed and an office was set up in St George's Hospital Medical School so as to be free from the constraints of an increasingly busy Editorial Office in Warwick Court. In spite of sterling efforts by the Editor the journal was not a success. It was launched with a subscription rate fixed on full economic costings with minimum circulation. Dr Rex Dawson, Chairman of the Publications Board at the time, writes: "Some of us believed that it should have been wedded to the *Biochemical Journal* for some time with it being given to the *BJ* subscribers as a free bonus for at least six months with the subscription rate adjusted to a level which would attract long-term subscribers. In fact the accountants won ...". It will be recalled that the ploy of giving *Biochemical Society Transactions* free for one year to *Biochemical Journal* subscribers paid off handsomely in the long run.

The number of original subscribers to *Bioscience Reports* were counted in the low hundreds and were obviously never in the foreseeable future going to reach 1000, the calculated break-even point. In spite of one or two attractive contributions in the form of manuscripts of Nobel Lectures, contributions were slow in arriving, and the financial loss was becoming too large to be justified as a service to the Society or even to Biochemistry in general. Eventually, on the recommendation of the Publications Board, the Committee decided to cut its losses and signed a contract with Plenum Press, who took over the copyright of the Journal for ten years on what could be considered as favourable terms for the Society. It was hoped that the back-up of a large organization with great experience in scientific publishing and particularly in promotion will allow the journal to establish itself as essential biological reading. The Board of Editors as well as the format will remain the same and the interests of the Society will be represented by the Publications Secretary, who at the time of the transfer was the late Dr G. B. Ansell (Fig. 6.13).

A combination of unpropitious circumstances conspired from the start to put the viability of the new journal at risk. The problems included (i) the overall international depression in science funding at the time of launch, which must have deterred potential new subscribers, (ii) the unexpected ability of the current journals to absorb most of the high quality copy coming forward and (iii) an expensive method of setting used in a laudable attempt to achieve rapid publication.

6.7 *Essays in Biochemistry*

One of the first proposals which the newly formed Advisory Committee for Publications had to consider in 1962–1963 was that an annual soft back *Essays in Biochemistry* should be published. The aim was to provide essays "which could be read with pleasure and profit by senior students and lecturers in Biochemistry. Each essay (would present) an overall view of one aspect of the subject, indicating its origin, present status and likely future development". A positive recommendation to the main Committee was accepted and in September 1963 Professor P. N. Campbell (Plate 1B) and the late Dr G. D. Greville were appointed editors. The launching of *Essays* was, however, not without incident. Dr M. G. MacFarlane, on being invited to provide a contribution for the first volume, replied by pointing out that in her opinion the Committee did not have the power to publish such a series without approval of a General Meeting of the Society. The rule (13) which she quoted specifically referred to publication of a "journal"; the Committee did not see *Essays* as a journal and the Symposia series was quoted as a precedent. However, it was agreed to take the proposal to the A.G.M. in September 1964. The proposal was carried (26–19; once again it is obvious that an important decision was made on a very small number of votes) and the first volume published by the Academic Press appeared in 1965; it was extremely successful: by 1968 over 7000 copies had been sold and in his Preface to the tenth volume Professor Campbell recorded that over 60,000 copies of the first nine volumes had been sold. The pricing policy agreed at the A.G.M. at which the project was approved, was that the volumes should be financially within the reach of students, whilst not losing money for the Society. This has been achieved throughout the existence of *Essays*, although sales have dropped markedly in recent years. Apart from the general recession in book sales this drop reflects once again the consequence of the highly specialized nature of modern Biochemistry. One cannot expect many Biochemistry students with an immediate interest in only one out of four essays

Fig. 6.13. Dr G. B. Ansell, Honorary Publications Secretary, 1980–1986 (deceased 21 November 1986).

buying a volume three-quarters of which is of no direct interest or quite frequently almost unintelligible to them. The guiding force of *Essays* for many years was Professor Campbell, who served from 1965 until 1985. The late Professor F. Dickens (Plate 1A) was a particularly effective co-editor from 1970, after Dr Greville's untimely death, until 1974.

An interesting aspect of *Essays* is that it has continued to be published by Academic Press although the Society has taken over many other publishing activities it has initiated. A possible change in publisher was considered in 1985 when Academic Press moved its London office to the U.S.A., but the Committee decided in December 1985 to continue in the same way, following some assurances for the future by the Press.

6.8 *Essays in Medical Biochemistry*

The proposal brought in 1970 before the Advisory Committee on Publications for the introduction of a new title in the *Essay* form, *Essays in Medical Biochemistry*, was eventually accepted after a working party reported positively, and the first volume appeared in 1974. However, the series was not as financially successful as expected in spite of good reviews. Increased biochemical specialization was again one of the reasons for the poor performance and, following poor support from subscribers, the Society reluctantly decided in 1979 to discontinue the venture with volume 4.

6.9 Biochemical Society Symposia

The events leading to the establishment of the Society Symposia and the decision to publish the proceedings of the meetings, together with their development to the present day, have already been described in Chapter 3.

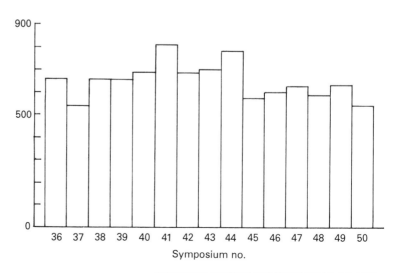

Fig. 6.14. Pattern of sales of Biochemical Society Symposia (nos. 36–50, 1973–1985) at the end of the first year of publication.

The publication of Symposia was in the early days in the hands of the C.U.P. but was transferred to Academic Press in 1964, when the formation was changed from softback to hardback. The Society took over full responsibility for their publication in 1971. Sales have always been satisfactory but never in the same league as *Essays in Biochemistry*. Recently, in common with all the Society's publications, sales are decreasing. Financially the series just manages to keep in balance (if overheads are disregarded) but, as with *Essays*, the series is considered to be mainly a service to members and not merely a money spinner. It will be remembered that shortened reports of *Group* Colloquia, as distinct from *Society* Symposia, are reported in *Transactions*.

The pattern of sales of Symposia over the years is recorded in Fig. 6.14.

6.10 *Biochemical Society Bulletin*

The origins of this publication have been described in Chapter 3. It is now well established as the Society's House Journal and currently each issue contains some 50 pages consisting *inter alia* of short articles of topical interest, Society news and meeting arrangements. A good indication of its scope is given by the contents list of a typical issue. Table 6.2 records such a list for the Bulletin of August 1985, which was issued for a meeting of the Society held in Belfast in September. The Abstracts of communications for a meeting are now contained in a separate booklet which is distributed with the appropriate issue of the *Bulletin*. Currently the Honorary Officers are making great efforts to improve and widen the general appeal of this membership publication.

6.11 Special (Occasional) Publications

Because of the difficulties of entering the book publishing sphere, outlined earlier in this chapter, the number of publications which can be considered booklets or books which the Society has published is small. A very strong special case has to be made before the Publications Board recommends publication, and even then it has to surmount the hurdles of the Finance Committee and Main Committee. The titles which have been issued over the years include: *Biochemistry, Molecular Biology and Biological Sciences*, a report of a sub-committee under Sir Hans Krebs set up to consider the Kendrew report (H.M.S.O. Comd. 3675) on Molecular Biology; *Safety in Biological Laboratories* (1978, reprinted), *Writing a Scientific Paper* (1979, reprinted) by V. H. Booth, the most successful of the Special Publications with over 15,000 of the first edition sold.

Conscious of the increasing importance of chemical education at school level the Society instigated the writing of a

Table 6.2. Contents of *Biochemical Society Bulletin* vol. 7, no. 3, 1985

Bulletin Articles	**The International Biochemical Scene**
	Federation of European Biochemical Societies
Editorial	
Why Public Relations?	**Belfast Meeting**
Scientific Procedures on Living Animals	Arrangements and Programme
Award Winners 1985	
The Society's Staff	**Classified Advertisements**
Biochemistry at Queen's University, Belfast	**Forthcoming Meetings**
FEBS Activities	Society Main Meetings
Krebs Memorial Scholarship 1985–86	Group Meetings
Members' Correspondence	Refresher Courses
Special Colloquia	Harden Conferences
Biochemical Society News	**Free Communications**
The Society's Regional Group Structure	Instructions to Authors
The Biochemical Journal	
	Diary of Events
The Society's 1985 Medals and Awards	
Grants and Fellowships	**Announcements**
Membership Subscription	Earthquake Damage to the University of Chile
News of Members	ISN/ASN Joint Meeting, Venezuela, 1987
New Members	
	Reservation/Registration Forms
	Belfast Meeting 24–27 September 1985
	London Meeting 18–20 December 1985
	Peptide and Protein Group Meeting
	Form of Nomination for Membership

Centre pull-out supplement: 615th Meeting (Belfast) Schedule

---

text-book for schools: *Introducing Biochemistry* by E. J. Wood and W. R. Pickering (the latter a practising school teacher). This was published commercially by John Murray in 1982 but its production was monitored academically and aided at all stages by the Society and can be, by stretching a point, considered a Special Publication of the Society.

Such a collaborative venture with a publisher was an important departure for the Society because for the first time the idea was introduced that its logo could be used as a seal of approval to promote a venture which was essentially financed elsewhere. This seal of approval has now been extended to a series of teaching discs for the BBC micro computer, published by IRL Ltd., but carefully monitored by the Society throughout their production.

It should be noted here that one video produced by the Society has already been published and that in order to encourage the making of films and videos in schools on life science subjects the Society now offers a series of awards at the Biennial National Schools Film and Video Festival.

6.12 The Future

The financial stability of the Society as we know it today, with elaborate free meetings, generous travel grants and a low annual member's subscription rate, is obviously based on the continuing success of the *Biochemical Journal*. There is no reason to believe that this situation will not obtain for some time to come but one is also aware of ongoing, vast technological (electronic) changes in the printing and communications industry. Soon authors may be asking to submit their papers on disc and will expect the editing and processing to be carried out electronically. This could mean that eventually a 'soft' version of the *Journal* would be available at the authors' own computer terminals. This development, unless prudently handled, would undoubtedly increase costs and reduce circulation and, probably, revenue. The Society Committee is well aware of such possibilities and its financial policy is designed to ensure long-term stability with appropriate investments (see Chapter 4).

References

1. Morton, R. A. (1969) *The Biochemical Society. Its History and Activities 1911–1969*. pp. 160. Biochemical Society, London.
2. Rogers, H. J. (1984) Letter to Author deposited in the Biochemical Society Archives.
3. Pogson, C. I. (1986) Biochemists' attitudes towards the *Biochemical Journal*: some conclusions. *Biochem. Soc. Bull.* **8** (1), 4–7.

Chapter 7

International Activities

7.1 Introduction — the First International Congress of Biochemistry
7.2 The International Union of Biochemistry (IUB)
7.3 Federation of European Biochemical Societies (FEBS)
7.4 Other International Activities

7.1 Introduction — the First International Congress of Biochemistry

Very soon after the Second World War it became apparent that Biochemistry was on the march and that effective international collaboration was necessary to achieve maximum rate of development of the subject.

The first post-War International Congress of Physiology was arranged for 1947 in Oxford and the Biochemical Society approached the organizers to ask them "to ensure that Biochemistry was allocated its share of the programme". The reply was that it was "impossible to issue a general invitation to biochemists to participate in the Congress and that while no actual embargo would be placed on biochemical papers, these would have to come from, or be introduced by, members of the Physiological Society". Even the mild R. A. Morton was moved to describe this as "a dusty answer" [1]. However, it served to stimulate the main Committee, in particular J. N. Davidson (Fig. 3.14), to start on the attempt to organize a Congress of Biochemistry with full international status, a difficult exercise because at that time there was no International Union of Biochemistry. In general, International Unions are the responsible agencies for organizing international congresses. The Committee of the Physiological Society responded by confirming their original stance but added that "if the Biochemical Society decided to initiate Congresses of their own they would have the Physiological Society's blessing, encouragement and offer of assistance". The project was also officially recognized by the International Union of Pure and Applied Chemistry (IUPAC) but without commitment for the future.

Although first informal surroundings did not reveal strong support for the idea the Committee was sufficiently convinced of its viability to send out letters to 100 selected biochemists in which they were invited to give their views on the need for an international congress. Only 24 replies were received but all were in favour, all approved of Cambridge as the location and all felt that August was an appropriate time of the year to hold the Congress. Ernest Baldwin, then at Cambridge, but later Professor of Biochemistry at UCL, reported that Cambridge could accommodate 780 people "including a few women" in August 1949. So the Committee got to work; they allocated £1700 [£17,000] to the Congress funds and set up a Congress Committee to make appropriate arrangements. The Congress was held from 19 to 25 August in glorious weather which revealed Cambridge at its very best. In all ways the Congress was completely successful as evidenced by the attendance of 1741, of whom 700 came from 32 different countries. As Professor A. C. Chibnall (Fig. 2.10 and Plate 1B), the President of the Congress, said in his opening speech: "As evidence of a world-wide interest in Bio-chemistry we could ask for no finer demonstration" [2].

For those, like the author, whose first International Congress it was, the experience was unforgettable. One still remembers the excitement not only of meeting legendary figures but also of being allocated Darwin's rooms in Christ's, not that the rooms were ideally situated when one contracted a dose of food poisoning! A Garden Party at St John's added a typically English dimension to the international occasion (Fig. 7.1).

In his opening speech Chibnall reported that an informal committee from different countries would consider how machinery for organizing future biochemical congresses could be established. Sir Charles Harington (Fig. 2.7), the Chairman of this committee, reported at the closing session of the Congress and submitted three resolutions. In short these were: (i) that the invitation of the Société de Chimie Biologique to hold the next International Congress in Paris in 1952 be accepted with gratitude; (ii) that an International Committee for Biochemistry be set up comprising 19 delegates from 14 countries with Harington as chairman (the U.K. representatives were J. N. Davidson and H. Raistrick, Professor of Biochemistry, London School of Hygiene and Tropical Medicine); (iii) that the committee should approach the International Union of Scientific Unions (ICSU) with a request for recognition as the international body representative of Biochemistry with a view to the formal constitution of an International Union of Biochemistry (IUB) as soon as possible. These resolutions were carried unanimously and thus the future of International Congresses of Biochemistry seemed assured and the mechanism for the formation of IUB had been set in motion. However, the actual birth of IUB was by no means straightforward and the ultimate success of the negotia-

Fig. 7.1. Garden Party at St John's College Cambridge during the First International Congress of Biochemistry, 1949.

tions depended to a great extent on the persistence and diplomacy of the members of the Biochemical Society involved. The negotiations lasted six years.

7.2 The International Union of Biochemistry (IUB)

Although IUPAC had officially recognized the first International Congress of Biochemistry, its rider "without commitment for the future" suggested that it was not altogether happy with the development. This was reflected by the resolution early in 1949 by the British National Committee of Chemistry that "the proposal for an International Union of Biochemistry would be better replaced by a proposal to establish a joint committee between the International Union of Biological Sciences and the International Union of Pure and Applied Chemistry which should be its mother union". The British National Committee for Biology rejected the idea of an IUB mainly, according to Davidson, "on the grounds that a multi-

Fig. 7.2. Sir Chales Dodds, F.R.S. Chairman of the Society Committee, 1951-1952.

plicity of unions was to be deplored" [1]. Perhaps more ominously in September 1949 IUPAC reconstituted itself into six sections, one of which was to be devoted to biological chemistry. The chairman of this section was Professor A. W. K. Tiselius (Sweden), who was also a member of the International Committee set up in Cambridge in August 1949.

The next important development was that Professor E. C. Dodds (Fig. 7.2), who had been nominated as a delegate to the IUPAC Congress to be held in New York in 1951 by the British National Committee for Chemistry, was invited by the main Committee to represent the interests of the Biochemical Society. It was a step which was to cause some unexpected difficulties. Meanwhile draft statutes of an IUB drawn up by Harington were approved by the main Committee, who invited The Royal Society (who would eventually be the adhering body, as with all other International Unions at that time) to set up a National Committee for Biochemistry.

An official application to establish the IUB was forwarded to ICSU for its consideration at its meeting in Washington in October 1951; F. Dickens (secretary of the International Committee; Plate 1A) and J. N. Davidson were to present the case drawn up in a memorandum prepared by Davidson, Dickens, Dodds and Harington.

Just before the Washington meeting Tiselius was elected President of IUPAC in September in New York, thus leaving the chairmanship of the Biological Chemistry section of the organization vacant. At short notice Dodds, who it will be recalled was representing the interests of the Biochemical Society, was invited to replace Tiselius; this he did with some misgivings and "only on condition that it was understood and minuted that he was in favour of an independent union of Biochemistry and that he would continue to further the cause of an independent union" [1]. The acceptance of this post, despite the conditions he attached to his agreement, led to disappointment and bitter criticism from some supporters of an IUB. It was felt that this greatly weakened the applicants' case and this was presumably further affected when Murray Luck (IUPAC) wrote to Harington inviting the International Committee to nominate five persons to fill vacancies on the Committee of the Biological Chemistry section. There also seemed to be a lack of interest in an IUB on the part of many American biochemists.

In the atmosphere prevailing it was not unexpected that the ICSU meeting at Washington deferred the consideration of the proposal to found an IUB for one year. However, the Society Committee was undaunted and in May 1952 it circulated a questionnaire on the proposed IUB to all members. As in all these types of questionnaire only about 25% of the membership bothered to reply but those who did were overwhelmingly in favour of the proposal. A rather smaller majority was in favour of establishing an interim working arrangement with IUPAC. At their June meeting the main Committee reaffirmed its commitment to the formation of the IUB, after hearing from Davidson that American opinion was now moving towards the idea of an autonomous union. They also agreed that they would not object to British representatives serving on the IUPAC Biochemistry Section Committee provided that they continued to press for an independent union.

The International Committee met for the third time during the 2nd International Congress of Biochemistry in Paris in 1952 with Davidson in the chair in the absence of Harington. They approved the stance taken the previous year in Washington (to hold out for an independent union) and then agreed to meet the Biological Chemistry Section Committee of IUPAC immediately after the International Committee meeting. Thanks to the staunch work of the chairman a stormy meeting closed with the International Committee holding its ground. Later in the same year the Executive Board of ICSU at its meeting in Amsterdam heard the case for the formation of an IUB presented by Davidson, Florkin (Belgium), Brand (U.S.A.) and Westenbrink (Holland). The Board were more sympathetic than they were in Washington; they came to no definite decision but, "as a result of unofficial advice proferred during the meeting, but outside it" [1], the IUB was established as a going concern independently of ICSU.

By 1953 an Interim Council had been set up and national membership was being considered. In the summer the fortunes of the IUB received a considerable boost when the Biological Chemistry Section of IUPAC meeting in Stockholm, with Dodds in the chair, gave definite support to the new Union [3]. Professor A. Neuberger (Plate 3A), who was a member of this Section at the time, stated that in the face of much opposition from the chemists Dodds stuck to his view favouring the establishment of the new Union [4]. Dodds' decision to take the chair was thus vindicated and the Biochemical Society Committee showed its appreciation by instructing its Honorary Secretary to thank him for his valuable work.

The next step was to set up a British National Committee for Biochemistry as the adhering body to IUB. Such a Committee usually comes under the aegis of The Royal Society, which, however, could not act before the IUB had been formally accepted by the General Assembly of ICSU; the next meeting of that body was, unfortunately, not until 1955. So in the meanwhile the Biochemical Society decided to act as the interim adhering body and set up a provisional National Committee. In January 1955 the Interim Council of IUB, now evolved into the Constitutive Assembly of IUB, held its first General Assembly in the University of London. Representatives from 12 countries (15 countries had indicated their wish to join) met under the chairmanship of Professor Marcel Florkin (Belgium), who had succeeded Sir Charles Harington. The statutes were presented and formally adopted and the first officers and council were elected [5]. The U.K. members of Council were Sir Rudolph Peters (Fig. 3.11), Sir Charles Harington and Professor R. H. S. Thompson (Plate 1B), who was also elected Secretary-General and who served in this post with distinction for nine years. A formal letter of application by IUB for adherence to ICSU was sent to their Secretary-General. By the time of the second General Assembly of IUB, held in Brussels in August 1955 at the time of the third International Congress of Biochemistry, five more countries had been admitted to the Union. Later that month the 7th General Assembly of ICSU met in Oslo and the application of IUB for adherence was confirmed and accepted.

So after a long, protracted and sometimes acrimonious battle the IUB emerged with full independent status. Its continuing success is known to everyone and this is not the place to recount it (see [6]). It must be clear, though, that its successful launch was due in great part to the efforts and the persistence of the Honorary Officers of the Society, in particular Davidson, in the early 1950s. Biochemistry in general owes them a considerable debt. In spite of Davidson's leading role he always considered the lobby as international. In a speech at the 50th Anniversary Dinner he said that after the 1952 meeting with ICSU (see above) the delegates lobbying for IUB came away feeling that their mission had failed but "the passionate pleading of a polylingual Belgian (Florkin), a forthright Dutchman (Westenbrink), an irrepressible American (Brand) and a taciturn Scot (Davidson) must have had some effect".

The first formal British National Committee for Biochemistry was set up by the adhering body, The Royal Society, with Sir Rudolph Peters as its first chairman; it first met in June 1956. A list of members who have served as chairmen of the National Committee is given in Table 7.1. The terms of reference of this Committee, as for all National Committees, are "to promote the branch of science in which they are concerned, more especially as regards international requirements, to nominate delegates to represent the U.K. at meetings

Table 7.1. Chairmen of the British National Committee for Biochemistry

1956	Sir Rudolph Peters, F.R.S.
1958	Sir Frank Young, F.R.S.
1964	Professor F. Dickens, F.R.S.
1967	Professor R. H. S. Thompson, C.B.E., F.R.S.
1970	Professor A. Neuberger, C.B.E., F.R.S.
1977	Professor T. W. Goodwin, C.B.E., F.R.S.
1982	Professor S. V. Perry, F.R.S.

of the IUB and to initiate proposals or questions for discussions at such meetings".

Over the years comments have been made questioning both the necessity for The Royal Society rather than the Biochemical Society to be the adhering body and also the membership of the National Committee. In spite of strong representations, particularly by Professor W. J. Whelan (Fig. 3.10), that The Royal Society was too formal and remote for easy communications with the Biochemical Society, The Royal Society has continued to be the adhering body to IUB. One of the most telling arguments for the *status quo* was that a change might jeopardize government funding of the National Committee. Certainly at the present time the remoteness of The Royal Society in this matter cannot be sustained. One important development which arose from the long arguments was that in 1967 the Council of The Royal Society approved a recommendation that the Chairman of the Biochemical Society should be *ex officio* a member of the British National Committee. So its present composition is six representatives of The Royal Society, one from the Association of Clinical Biochemists, three from the Biochemical Society, one from the British Biophysical Society, one from the Nutrition Society, one from the Physiological Society, two from the Royal Society of Chemistry, one from the Society of Chemical Industry, one from the Society for Experimental Biology, one from the Society for General Microbiology and, *ex officio*, an officer of The Royal Society (at present the Biological Secretary) and the Chairman of the Committee of the Biochemical Society. Of the present members of the National Committee only three out of twenty are not also members of the Biochemical Society. So the voice of the Society is strongly heard on the National Committee at the present time and it would be surprising if the Society's views on most issues did not prevail.

7.3 Federation of European Biochemical Societies

As Europe was gradually recovering from the devastation of the Second World War and as travel and contacts became easier it was not surprising that the main Committee of the Biochemical Society turned its sights on joint meetings with their European neighbours. An early, probably premature, proposal for a meeting in Ghent in 1948 had to be dropped because of lack of support. However, a joint meeting at Oxford in 1956 with the newly formed Belgian Biochemical Society was highly successful. Meetings in Continental Europe were then organized by Professor P. N. Campbell (Plate 1B), the Honorary Secretary, after he had persuaded the Wellcome Trust to provide travel funds; the venues were Turku (1959), Paris (1960) and Louvain (1962). He also introduced the idea of inviting Continental European biochemists to the summer

Fig. 7.3. Professor F. C. Happold. First Chairman of FEBS, 1964. Chairman of the Society Committee, 1963–1965.

meeting of the Society held alternately at Oxford and Cambridge and Professor W. J. Whelan (Honorary Meetings Secretary) arranged such meetings in Cambridge in 1962 and in Oxford in 1961 and 1963, when they ceased for reasons which will soon become apparent. Meetings of British biochemists with their counterparts in Continental Europe, however, continued until 1965; the locations were Leyden (1963) and Santa Marghareta (1965). However, the Campbell/Whelan partnership, ably aided by Professor H. R. V. Arnstein (Plate 3A) (Honorary Meetings Secretary), and Dr W. F. J. Cuthbertson (Fig. 4.3) (Treasurer) had already had the idea of developing some form of association of Biochemical Societies in Europe and took the opportunity afforded by the Oxbridge meetings to explore the possibility informally with officials of the European Societies. Meanwhile in 1962 Whelan had resigned as Meetings Secretary to be reappointed in 1964 to a new honorary post very necessary because of these developments — International Secretary. In the meanwhile H. R. V. Arnstein had become Honorary Meetings Secretary. After a considered discussion in 1962 it was decided that a meeting of delegates from all European Societies should be called for the Oxford meeting in 1963. At this meeting it was proposed that a Federation of European Biochemical Societies (FEBS) be set up. Provisional statutes had been prepared by Whelan: "These had very simple aims. They provided in the main for the Societies to engage in mutual collaboration and, in particular, to hold a combined meeting somewhere in Europe every one or two years. In brief the proposals and statutes were accepted and the Federation was launched from 1 January 1964. It was decided to hold the first Federation meeting in London at UCL in March 1964 [7]". Professor F. C. Happold (Fig. 7.3), then Chairman of the Society, became the first chairman of FEBS, Whelan acted as Secretary of the meeting and Professor S. P. Datta (Fig. 7.4) as Treasurer. The meeting was clearly a success and amongst other innovations was the organization of a Trade Fair by Dr D. C. Watts. FEBS is now a household word amongst biochemists and Whelan [8] recalls that he has "still to learn any objections from Boots Pure Drug Company, who, I only discovered later, were already using 'FEBS' to describe a patent analgesic".

At the next meeting of FEBS in Vienna in 1965 it was decided that a Secretary-General and a Treasurer were necessary in spite of efforts to keep administrative activity to a minimum. Whelan was appointed Secretary-General and Datta Treasurer. When Whelan left for the U.S. in 1967, Arnstein was chosen to succeed him, an appointment which Arnstein says was "almost entirely due to the esteem in which the Biochemical Society was held at that time by our European friends ..." [9]. He had also, of course, considerable experience in running the Biochemical Society's affairs. Up to the present

Fig. 7.4. Professor S. P. Datta. Treasurer of FEBS, since 1964.

time Campbell has served FEBS in many capacities, Datta continues to be a most effective Treasurer and with the author, who, at that time was Chairman of the Publications Committee, recently concluded new and favourable contracts for the publication of the *European Journal of Biochemistry* and *FEBS Letters*. Datta was also an outstanding Managing Editor of *FEBS Letters* from its inception in 1967 until 1985, when he retired.

In 1981 FEBS returned to the U.K. to hold its 14th Congress in Edinburgh. The burden of organizing the Congress was carried out by a Committee consisting mainly of the Professors of Biochemistry in Scotland with the help of a full time Executive Officer; the Chairman of the Committee was Professor H. M. Keir (Plate 1B). The basis of the Congress was a series of 39 one-day Symposia so arranged by careful planning to avoid as far as possible overlap of cognate topics and thus to cause a minimum of distress to participants. Each Symposium was self-contained, consisting of two lecture sessions (invited speakers), one Poster Session and one Workshop Session. The abstracts of the meeting were published as a special issue of *Biochemical Society Transactions*. The Congress was a success both scientifically and financially and, not unexpectedly, the hospitality was of a high order.

Thus, as with the formation of IUB the Biochemical Society can congratulate itself that the efforts of Whelan, Campbell & Co. led directly to the setting up of FEBS. Without their drive, enthusiasm and belief in the internationalism of Biochemistry the Federation would, at worst, never have been founded or, at best, its inception would have been greatly delayed.

When Whelan became Secretary-General of FEBS, Professor A. P. Mathias (Plate 1B) succeeded him as International Secretary. With IUB and FEBS working smoothly this post rapidly became redundant and it was abandoned when Mathias followed Professor A. N. Davison (Plate 1B) as Honorary Meetings Secretary at the time the latter became Honorary Secretary. Recently pressure is mounting for the restoration of this office (see Chapter 3).

7.4 Other International Activities

In July 1980 the Society welcomed a delegation from Japan to its Sheffield meeting. The main object of the meeting was to honour Sir Hans Krebs, with a Colloquium to celebrate his eightieth birthday. Two of the visitors took part in this two-day programme. The third day of the meeting was devoted to a joint Biochemical Society/Japanese Biochemical Society Colloquium on "Proteins and Peptides" organized by the Peptide and Protein Group. The Japanese Society paid the travel expenses of their delegation and the British Council

Fig. 7.5. Wall hanging presented to the Society by the Chinese Delegation attending the Oxford Meeting in 1982.

covered their accommodation costs during their stay in the U.K. The Society, in its turn, sent eight delegates to attend a very large meeting of the Japanese Biochemical Society in October 1982. Apparently 4600 members, out of a total membership of 9000, attended the meeting. The U.K. delegates reported that their hosts' hospitality was on an equally generous scale.

In 1982 the Society also welcomed a delegation of Chinese biochemists to the Oxford meeting, just three years after the Chinese Biochemical Society had been admitted to IUB. As a souvenir of their visit the Chinese delegation presented the Society with a wall hanging (Fig. 7.5) which now graces the Committee Room at Warwick Court. The visit of the Chinese delegation reminded Professor L. Young (Plate 4D) that one of his Ph.D. students who granduated in 1942, Professor Zhang Longxiang, became president of Peking University.

The Society is also involved in the European Federation of Biotechnology; this is discussed in Chapter 5.

References

1. Morton, R. A. (1969) *The Biochemical Society: its History and Activities 1911-1969*. 160 pp. Biochemical Society, London.
2. Report of First International Congress of Biochemistry 1950. 50pp. Biochemical Society, London.
3. Dickens, F. (1975) E. C. Dodds, *Biogr. Mem. Fellows R. Soc.*, **21**, 227.
4. Neuberger, A. (1985) Personal Communication.
5. Thompson, R. H. S. (1963) *ICSU Review*, **5**, 142-146.
6. Thompson, R. H. S. (1978) I.U.B. Jubilee. *TIBS*, **3**, N63-N64.
7. Campbell, P. N. (1964) First meeting of the Federation of European Biochemical Societies. *Chem. Ind. (London)*, no. 23 (6 June), 961-965.
8. Whelan, W. H. (1969) Submission to R. A. Morton deposited in the Biochemical Society's Archives.
9. Arnstein, H. R. V. (1984) Letter to the author deposited in the Biochemical Society's Archives.

Chapter 8

Professional and Educational Activities

8.1   Introduction
8.2   Careers for Biochemists
8.3   Heads of Departments Conferences
8.4   Refresher Courses
8.5   Public Relations
8.6   Presence at FEBS and IUB Meetings
8.7   75th Anniversary Celebrations
8.8   Education
8.9   Animal Welfare
8.10 Postscript

8.1   Introduction

The Biochemical Society is not a professional body in the same sense as, for example, the Royal Society of Chemistry, and certainly at the moment has no ambitions in that direction. The relatively recent decision not to apply for a Royal Charter (Chapter 3) emphasizes this attitude. However, gradually the main Committee of the Society has moved more and more in the direction of providing additional services to its members, so that by March 1970 a Professional Sub-Committee with six members, one the Chairman of the Committee *ex officio*, was set up under the chairmanship of Dr G. A. Snow (ICI Pharmaceuticals) (Fig. 6.7) with no specific remit other than to look into general problems. The terms of office of members were to be three years with the possibility of re-election.

Over the next few formative years changes in its constitution were frequent. In April 1973 the Secretary of the Industrial Biochemistry Group became an *ex officio* member in place of an ordinary member; then in March 1975 its constitution was substantially changed. It was agreed that ordinary members should be elected each year and serve for three years; that the additional nominated member should come from amongst members of the Committee; that the Chairman should be elected by the Sub-Committee and would serve for three years from the date of his election as Chairman; that the Honorary Careers Adviser should be an *ex officio* member. In July 1977 the title of the Sub-Committee was changed to Professional

Fig. 8.1. Professor H. Baum. Chairman of PEC. First Public Relations Official.

and Educational Sub-Committee (PESC), thus reflecting the Society's increasing concern with the training of biochemists. Further, the Sub-Committee was given an annual budget of £1000 and, for the first time, detailed terms of reference: responsibility for promotion and planning in respect of the development of Biochemistry as a profession; initiating and co-ordinating the Society's activities in the field of education; reviewing the supply, demand for, and training of biochemists; keeping under review all major issues relevant to the above, and as referred by the Committee, and making recommendations. These Terms were extended in November 1979 to include advising on Public Relations. The holder of the newly established post of Public Relations Official (PRO) was added to the Committee as an *ex officio* member. The first PRO to be appointed was Professor H. Baum (Fig. 8.1). In 1985 the increasing importance of the Committee was recognized by elevating it to a Board of the main Committee, the Professional and Educational Committee (PEC), thus putting it on an organizational level with the Finance Board and Publications Board. The Chairmen of PE(S)C since it was founded are given in Table 8.1. The various major professional aspects of Biochemistry to which the PE(S)C have so far applied their minds will be discussed in the following sections, but they are continually widening their net: the Agenda for a recent meeting contained some 75 items! Many activities which now are well established aspects of the Society's business were considered only spasmodically before the existence of PE(S)C.

Table 8.1. Chairmen of the Professional and Educational Committee*

Dates	
1970–76	G. A. Snow
1976–77	G. Boyd
1977–78	P. J. Heald
1978–81	F. W. Hemming
1981–84	H. R. V. Arnstein
1984–	H. Baum

*A Sub-Committee until 1985.

8.2 Careers for Biochemists

The first interest the Society showed in this problem was in 1960 when they invited the late K. S. Dodgson (Fig. 8.2) to expand a booklet which he had written for his own Department at University College, Cardiff, so that it would be suitable for national distribution. Dr D. S. Jones (Fig. 8.3), currently Careers Adviser, takes up the story:

"It is interesting to note that when the booklet was written only 15 universities including two London Colleges, offered first

Fig. 8.2. Professor K. S. Dodgson. Honorary Secretary, 1964–1969. Chairman of the Society Committee, 1983–1986.

degree single honours courses in Biochemistry. (Today about 50 universities and colleges offer such courses.) Looking through that booklet there are many items which in the light of the situation today make interesting comparisons. For example, referring to positions in hospitals as biochemists (today's basic grade biochemist) it was stated that a degree in Biochemistry is not essential. I wonder what today's graduates — and Ph.Ds even — think of that comment in the light of the current competition for such positions.

"The booklet has been revised and reprinted on several occasions since. In September 1962 the first revision was carried out by Professor Dodgson and then in September 1968 a new edition was produced by Dr (now Professor) Gillian M. Powell of the University of Wales, University College, Cardiff. The edition expanded the original booklet from 8 pages to 30 pages and by this time 25 universities were offering single honours courses in Biochemistry with several joint honours courses also being offered. Still in 1968 it was not considered essential to hold a degree to gain a post as a hospital basic grade biochemist. The next revision by Professor Powell was in 1975 and at this point the title of the booklet was changed from *Careers in Biochemistry* to *Careers for Biochemists*. This change was significant because, by the mid-1970s, with the increase in the number of Biochemistry graduates and the general employment situation in the U.K. starting to decline, it was clear that there was not employment in Biochemistry for all Biochemistry graduates. The revision therefore included a section which pointed out the opportunities for Biochemistry graduates outside the Biochemistry field, e.g. in management, finance and administration. For the first time photographs of people at work in Biochemistry laboratories were included in this edition.

"A further revision was completed in 1979, this time by Mr P. D. Deary (Careers and Appointments Service) and Dr D. S. Jones (Biochemistry Department), both of the University of Liverpool. At this time because of the rapid increase in inflation and the corresponding rise in salaries, information on salaries was inserted as a separate leaflet and this has been revised on a yearly basis. As well as illustrative photographs, this edition also contained a few cartoons which it was hoped would have an appeal to sixth-formers in schools. The booklet has been revised again by the same authors and published in 1986."

Fig. 8.3. Dr D. S. Jones. Honorary Careers Adviser, since 1978.

The early editions of this booklet were aimed at potential students of Biochemistry, but in 1970, when PESC was set up, little attention had been paid to the career prospects of new graduates. This was no doubt due to biochemists being very much in a sellers' market, but by 1973 the writing was on the wall and the main Committee established the position of Honorary Careers Adviser, who would be answerable to PESC and would be an *ex officio* member of that Committee. The first holder of this office was Professor Gillian Powell (Cardiff, Fig. 8.4), who covered the period 1973-1977. She found that she was dealing with queries about careers and education from individuals, careers organizations and career

Fig. 8.4. Professor G. M. Powell. First Honorary Careers Adviser, 1973-1977.

advisers and with requests for articles from various publications concerned with careers. Professor H. M. Keir (Plate 1B; Adviser from 1977 to 1979) extended the activities by developing relations with the Royal Society of Chemistry, the Society for General Microbiology and the Institute of Biology. Since 1978 Dr D. S. Jones has been the Society's Honorary Careers Adviser and, in a period which has seen serious difficulties arise for the first time in the employment of biochemists, has recently been making strenuous efforts to improve the liaison between industry and the Society by visiting a number of organizations which are potential employers of trained biochemists.

8.2.1 Employment Surveys

In 1970 PESC found that little attention had been paid to the career prospects of young graduates and when the Sub-Committee set about accumulating information they found that surprisingly little was available. Dr Snow recalls:

> "University departments were reticent about disclosing their examination results or of providing precise information about the jobs to which their graduates went. The Sub-Committee decided to send out a questionnaire on these matters to all biochemical departments in the British Isles. They undertook that only collective results would be published; individual sources would not be identified. The response was better than expected and the usefulness of the data was quickly appreciated. Methods were refined in the light of experience and the yearly survey has become an accepted feature of the Society's activities. Its value has increased with time, since it can detect trends and so provide a guide for the future. In particular it provided a measure of the balance between the number of graduates being produced and the job opportunities available to them. In some branches of science numbers of graduates have greatly exceeded any likely outlet, with resulting waste and frustration. This has not happened in Biochemistry, but the surveys showed that the danger was narrowly avoided."

Professor F. W. Hemming, Chairman of PESC (1978-1981; Fig. 8.5) continued the annual surveys until Dr D. S. Jones, the present Honorary Careers Adviser, took on the job. The questionnaire sent out to Departments has developed considerably and information is collected on: the movement of scientists between different disciplines in universities posts; the age profile of academic staff in University Biochemistry Departments; the entry of Biochemistry graduates into careers in Biotechnology; and statistics on the application and admissions of students into Biochemistry Courses. The comprehensive approach to the problem is well illustrated by Table 8.2, taken from the 1985 report of the Honorary Careers Adviser. Dr Jones has summarized the trends observed since the survey began in 1970:

Fig. 8.5. Professor F. W. Hemming. Chairman of PESC, 1978-1981.

Table 8.2. Entry of biochemists of all qualifications into classes of employment

Originally published in *Biochemical Society Bulletin*.

	Category	1984 No.	1984 (%)	1983 No.	1983 (%)	Av. 1978–82 No.	Av. 1978–82 (%)	*	
A1	Further biochemical training	450	(26.5)	414	(25.4)	435	(26.2)	S†	↑
A2	P.G.C.E.	69	(4.1)	N/A		N/A			
B1	Univ. staff, permanent	8	(0.5)	17	(1.0)	35	(2.1)	V	↓
B2	Univ. staff, temporary	35	(2.1)	29	(1.8)			V	↑
B3	Polytechnic staff	4	(0.2)	2	(0.1)	4	(0.2)	S	↑
C	Research institutes	57	(3.4)	45	(2.8)	45	(2.7)	V	↑
D	Univ./poly. technicians	58	(3.4)	47	(2.9)	60	(3.6)	↓	↑
E	Civil Service and Public Authorities	14	(0.8)	21	(1.3)	26	(1.6)	↓	↓
F	Hospital laboratories	71	(4.2)	66	(4.0)	94	(5.7)	↓	↑
G	Industry	112	(6.6)	110	(6.7)	124	(7.5)	↓	S
I	School teaching	8	(0.5)	79‡	(4.8)	79‡	(4.8)	V	S‡
	Biochemical employment	367	(21.6)	416‡	(25.5)	467‡	(28.1)	↓	S‡
P	Further other studies	159	(9.4)	158	(9.7)	163	(9.8)	S	S
M	Commercial/sec. work	116	(6.8)	108	(6.6)	117	(7.0)	V	↑
N	Misc. employment	42	(2.5)	40	(2.5)	31	(1.9)	↑	S
S	Other classification	1	(0.1)	1	(0.1)	15	(0.9)	S	↓
	Non biochemical employment	159	(9.4)	149	(9.1)	163	(9.8)	S	↑
H	British → Abroad (train.)	52	(3.1)	72	(4.4)	55	(3.3)	S	↓
J	British → Abroad (employ.)	25	(1.5)	31	(1.9)	19	(1.1)	V	↓
K	Non-British → Abroad	120	(7.1)	88	(5.4)	113	(6.8)	V	↑
Q1	Unplaced, seeking	121	(7.1)	177	(10.9)	125	(7.5)	↑	↓
Q2	Unplaced, not seeking	21	(1.2)	24	(1.5)	22	(1.3)	↑	S
R	Unknown	154	(9.1)	102	(6.3)	103	(6.2)	V	↑
	Total	1697		1631		1660		S	↑

* Trend over past few years, change 1983 → 84.
† S = Steady; V = variable; ↑ = increase; ↓ = decrease.
‡ Includes Postgraduate Certificate of Education (P.G.C.E.).

"The employment surveys have shown that there has been a steady increase in the number of biochemists graduating throughout the 1970s, reaching a peak in 1981. Since then the output seems to have more or less steadied. Of the total of biochemists at all levels of qualification coming on to the employment market, whereas in the early 1970s about 64% remained in a Biochemistry related area [either in further training (33%) or in employment (31%)], in 1984 53% did so (27% in further training and 26% in employment). In the early 1970s 6% entered non-biochemical employment whereas in 1984 this has risen to 9%. (These are mainly first degree graduates.) Also in the early 1970s 5% of biochemists were unplaced at the time of the survey whereas in 1984, 8% were unplaced."

8.2.2 Regional Careers Conferences

As a logical development from the Employment Survey PESC organized Regional Careers Conferences in 1970–1971. They were held in the Universities of Birmingham, Glasgow, Liverpool and London. In spite of the efforts put in by the local organizers and speakers from organizations employing biochemists the conferences, although useful, were, except in the case of that held in Glasgow, not outstandingly successful and were not continued. They did, however, alert Heads of Departments to the difficulties ahead and the need to provide students with more information about careers.

During the next decade the need for a resumption of the Careers Conferences was becoming more and more urgent. They were restarted in 1980 and since then Dr Jones has organized a number of successful regional conferences. The first was held in Leeds in 1980

> "... for students in Biochemistry Departments from North of England Universities. As well as speakers other representatives from industry were invited and the programme included adequate periods of time for informal discussions between the industrial representatives and the students. Industry was asked to support the conference financially and through the generosity of many firms the conference was self-financing. The Leeds conference proved to be very successful with an attendance of approximately 200 students. Since that time conferences have been held as follows: 1981, London; 1982, Bristol and Edinburgh; 1983, Leeds and Dublin; 1984, London and Glasgow; 1985, Leeds and Bristol [and 1986, London and St Andrews]. Each conference has been styled in a similar fashion to the one held in Leeds and industry has continued to support them financially and in sending representatives. During the past three years the Society has also supported the conferences financially. They have proved very popular with students such that at some of the conferences a limit has had to be placed on the number of students from any one department being allowed to attend."

8.2.3 Other Career Activities

In addition to the major activities just discussed the Society has recently been trying to make its presence felt in careers conferences organized by schools or education authorities. This is usually achieved at the local level by putting schools in touch with the Biochemistry department of their nearest university. Currently a pamphlet is being prepared for distribution at Schools Career Conferences.

On two occasions (1971 and 1981) the Society has attempted to set up an employment register to facilitate contact between prospective employers and employees; on neither occasion did the scheme receive enthusiastic support.

8.3 Heads of Departments Conferences

The first meeting of Heads of Departments sponsored by the Society was called in 1967 by Professor G. R. Tristram. In

1970 one of the first decisions PESC took was to reintroduce the Conference and to make it an annual event. The first new style meeting, held at the A.G.M. in London in 1971 with Dr Snow in the chair, was reasonably successful. It was sufficiently successful for the decision to be taken to make it an annual event. Dr Snow recalls rather wryly: "They proved something of a nightmare to the chairman. The meeting was always reluctantly interpolated into the already overcrowded schedule of the Annual Meeting. It thus tended to occur at an awkward time conflicting with lunch or an important scientific meeting, and was often relegated to an unsuitable or crowded room where people could not easily see or hear one another. In spite of these difficulties the value of the meetings was recognized and gradually they developed a more regular and effective form".

This is certainly true and nowadays the arrangements for the Conference are much more structured; time is specifically set aside in the programme of the meeting selected; clashes with scientific activities are thus avoided. Reports of the Conferences are published annually in the *Bulletin*. One significant outcome is that Heads of Departments can frequently speak authoritatively with one voice on many important issues. This makes their *impact* on higher bodies much greater than previously. Whether in these days the *effect* on higher bodies is greater remains debatable.

8.4 Refresher Courses

One of the first problems the PESC addressed was the pace at which Biochemistry was advancing. It was suggested that occasional colloquia should be organized to keep biochemists up to date on developments in rapidly expanding fields. Such Refresher Courses should be run by individual university departments and made self-supporting by charging an appropriate fee. The first refresher course, proposed by the late Professor G. Boyd (Fig. 8.6), was held in the University of Leeds and organized by Professor P. N. Campbell. The subject was "Nucleic Acids and Protein Biosynthesis". Four days were allocated to the course, which consisted of 15 lectures; the course fee was set at £15, which covered all organizational expenses and honoraria to speakers of £15–20. Sixty attended, most of whom were from Polytechnics and Industry. A loss of £133 was easily covered by the Committee guarantee to underwrite the first course by up to £400.

Following this a number of courses were proposed; some had to be cancelled because of lack of support, some did not materialize, but the majority were scientifically successful. Early successes were courses on Enzymology (1973), Physical Techniques (1974, oversubscribed) and Chloroplasts (1975). Surprisingly the 1974 course on mitochondria had to be

Fig. 8.6 Professor G. Boyd. Chairman of PESC, 1976–1977.

cancelled owing to lack of support. The fees, greater for non-members than for members, were fixed so as to provide a small but significant surplus. With the exception of one or two near disasters, this aim was achieved and the profits were shared equally between the Society and organizing departments. As time went on Society funds increased whereas those of University Departments decreased, so that in 1982, the share of the surpluses was changed: one-third to the Society and two-thirds to the organizing department.

In 1984 in order to make it easier for graduate students to attend Refresher Courses ten bursaries of £75 each were made available each year.

The topics, organizers, attendances and venues of some recent Refresher Courses are recorded in Table 8.3.

8.5 Public Relations

In 1976 positive steps were taken by the main Committee, on the recommendation of the Publications Board, that a Promotions Organizer should be appointed to explore ways of increasing the sale of the Society's publications and of generally publicizing the Society's activities. Dr Snow took on this difficult job and his main activities were concerned with design of material for advertising publications, publicity for the Society and study of factors affecting the circulation of the *Biochemical Journal*. His efforts on behalf of the *Biochemical*

Table 8.3. Refresher Courses organized since 1950

Date	Title	Organizer	Location	Attendance
9.80*	Biochemical Basis of Human Disease	J. R. Griffiths/ J. Hermon-Taylor	St. George's HMS, London	—
9.80	Immunoassay	G. S. Challaud	St. Bart's. HMC, London	36
4.81	Xenobiotics	D. V. Parke/G. G. Gibson	Surrey	32
4.81	Techniques in Intermediary Metabolism	C. I. Pogson	Manchester	20
10.81	Cellular Immunology	J. Taverne/H. M. Dockrell	Middlesex HMS, London	47
9.82	Microcomputer/Microprocessor	R. E. Dale	Manchester	13
9.82*	Glycoproteins	R. D. Marshall	Strathclyde, Glasgow	—
7.83	Biochemistry of the Nervous System	A. N. Davison	Institute of Neurology	60
9.83	Animal Cell Culture	R. Tindle *et al.*	Beatson Institute, Glasgow	30
9.84	Recombinant DNA	G. E. Blair	Leeds	65
9.84	Hormone Receptors	D. Schulster	Middlesex HMS, London	38
12.84	Development & Application of Bioelectrodes	C. R. Lowe	Cambridge	47
3.85	Current HPLC Practice for Biochemists	R. W. A. Oliver	Cardiff	59
4.85	Free Radicals in Biochemistry	D. V. Parke	Surrey	51
3.86	Subcellular Structure and Function	T. J. Peters	Clinical Research Centre, Harrow	32
4.86	Mass Spectroscopy in Biochemistry	R. W. A. Oliver/ J. S. Thompson	Liverpool	20
9.86	Nucleic Acid Synthesis, Sequencing and Function	A. D. B. Malcolm/ L. C. Archard	Charing Cross & Westminster MS	43

*Cancelled because of lack of support.

Journal have been summarized in Chapter 6, and the story of the Society logos in Chapter 3. With regard to advertising material it was decided that individual leaflets for the Society's publications, other than journals, should be left with the publishers concerned: "The printers have staff capable of producing acceptable, if not very exciting designs and the work is probably best left with them. In any case the numbers sold are too small to yield the Society much profit; indeed they often have to be subsidized". Dr Snow continued: "... It was regarded as more important to explore ways of publicizing the *Biochemical Journal* and *Transactions*. Accordingly a brochure was devised. This had a striking cover featuring a space satellite circling the Earth, symbolic of the rapid advance of Biochemistry. Inside, the particular merits of the Society's periodicals were listed and examples were given of important recent papers appearing in them. These brochures were used to hand out at international biochemical gatherings, meetings in the U.S.A. of the Special Libraries Association, etc. and were sent out to selected institutions. They attracted comment, mostly favourable, but it was impossible to determine whether they had any impact on sales. A second, updated version was produced; they were then discontinued. On balance, the cost and effort of production did not seem warranted by the results".

Dr Snow's somewhat pessimistic view of all his promotional activities hinted at in the last sentence may be justified, but it is certain that all his hard work clarified many issues and provided a firm foundation for later activities, particularly in dealing with journal promotion.

By 1981 when Dr Snow retired as Promotions Organizer the situation was changing rapidly and Biochemistry in general with other sciences was coming under considerable pressure, not only financial, and the then chairman of PESC (Professor H. R. V. Arnstein; Plate 3A) felt that there was an urgency to promote the image of Biochemistry in the face of increasingly unsympathetic public reaction to science and education.

Even the connection of Biochemistry with Biotechnology, which had the respectability of government approval, did not ameliorate the situation significantly. To encourage public awareness of the need for strong support of research and training in Biochemistry the PESC recommended that a position of Honorary Public Relations Official be established. This was eventually accepted by the main Committee and, as indicated earlier in this chapter, in 1982 Professor H. Baum was the first appointee. He currently pursues this activity with energy and flair. To help with this aspect of the Society's affairs and with the increasing information requirements of other expanding activities of the Society, a Research and Information Officer was appointed in 1985 (see Chapter 3). One of the major duties of this Officer is to service the PEC.

Developments in public relations have included: the establishment of press releases to scientific and national press as a matter of routine on Society meetings; the provision of information on the Society to MPs, Parliamentary Select Committees and science writers for radio and television; provision of a list of Society members who would act as official spokesmen on urgent and topical matters to MPs, the media, appropriate members of the House of Lords and to the recently formed CIBA Foundation Media Service. Indeed many of these activities foreshadowed The Royal Society's report on "the Public Understanding of Science". The overall effect of these activities has certainly been favourable, with one or two real achievements, such as in 1986 when the Chairman of PEC was invited to write an editorial on the Society's response to the Government's Green Paper on Education for the *Journal of the Royal Society of Medicine*. The article was subsequently published in the *Bulletin* [1]. Some disappointments have also occurred, as when in 1984 a seminar proposed for the Association of British Science Writers was dropped because of lack of support.

The need for improved public relations within the Society itself has been acknowledged recently by reports and extended articles on the Committee's activities as well as on other relevant topics in the *Bulletin*. Important topics which have engaged the PEC recently are the 75th Anniversary Celebrations (see below) and, closely related to the celebrations, promotion of a permanent Biochemistry Gallery in the Science Museum in South Kensington. This turned out to be too ambitious in the time available but a smaller introductory exhibition, organized in collaboration with the Society by Dr H. Kamminga and entitled "Cells, Molecules and Life", was mounted in time for the December 1986 Meeting (see also page 76).

8.6 Presence at FEBS and IUB Meetings

The presence of a Society stand at IUB and FEBS meetings began at the FEBS meeting in Paris (July 1975) when Doris Herriott (Meetings Officer, Plate 2A) set up a small display of Society material. This also acted as a base for her general activities at the meeting. Although then this activity apparently upset the organizers the provision of a stand has become a regular feature of FEBS and IUB meetings. Since 1969 the organization of the stands has been in the hands of Dr D. C. Watts and he has been greatly helped more recently by Dr Elizabeth Evans, who organized the very successful mobile exhibition in connection with the 75th Anniversary Celebrations (see Section 8.7). The very professional stands of recent Congresses are exemplified by that seen in Fig. 8.7 (Perth, Australia, 1982). Dr Watts has given his version of the way this activity developed, in his own inimitable style:

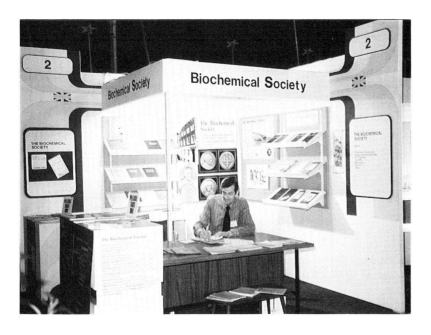

Fig. 8.7 The Society's stand at the IUB Congress in Perth, Western Australia, 1982. Dr D. C. Watts is seen in attendance.

"I have organized the stand at Toronto (1979) and subsequently Jerusalem, Edinburgh, Perth (where the display material never arrived due to strike action and I had to recreate the stand in one weekend before departure), and Brussels. There was no stand in Moscow. The attitude of the officers generally has, for the most part, been one of benign indifference so long as I did not spend too much of the Society's money! Most of the display material has been produced by me, or the art department here at Guy's, on a shoestring (and, you might say, looked it!) but it received a favourable, even envious, response from the members of other societies. The average cost to the Society for display material for the whole stand has been under £200 on each occasion. The commercial production of one poster would cost that. The *BJ* eventually became interested in the Society's stands and the Brussels meeting saw two panels contributed by Elizabeth Evans promoting the *BJ* (I had actually previously had a simple poster advertising the *BJ*). They cost as much as the rest put together, but never mind, it was the first real show of practical interest from anyone else at all (other than Doris Herriott, who has always attended these meetings and likes the stand as a base). This year [1986], for Amsterdam, Elizabeth is designing the stand in collaboration with myself, and she will organize all or most of the artwork. The cost will go up but the Society is not now hard up and at last recognizes that the stand does fulfil a useful publicity function even if it cannot be evaluated in hard commercial terms.

"In connection with the stand I organized a few give-aways, pens with the Society logo etc. It was because of this that when the Society suddenly decided it wanted a tie I was given the job of organizing it. It was not my original idea, however — at least on this occasion. The give-away situation can become quite delicate. I currently have a stock of Society keyrings, bought when give-aways were in favour; before they could be used they went out of favour! I wait for them to come into favour again."

It is interesting to note that the keyrings came into favour quite quickly and at the IUB Congress in Amsterdam in 1985 they sold "for a modest sum, like hot cakes".

The latest stand at the FEBS meeting in Berlin was entirely under the control of Dr E. Evans, and involved for the first time the exhibition of the Society's software (with IRL Ltd.) and videos.

8.7 75th Anniversary Celebrations

In 1983 the main Committee set up a Working Party to organize a programme to celebrate the Society's 75th Anniversary. The activities which were masterminded by PEC, generally fell into two categories, one demonstrating the justifiable pride of the Society in its achievements over the years and the other exploiting the opportunity for effective public relations.

In the first category a mobile exhibition has been mounted based mainly on the Society's development over 75 years. It was prepared by Dr E. Evans (Fig. 8.8), who has also been concerned with the Society's stands at recent European and International Congresses. She has the benefit of advice from Professor G. Barker (Plate 2C), the present Society Archivist. Part of the mobile exhibition is illustrated in Figs. 8.8 and 8.9.

Fig. 8.8. Part of the Mobile Exhibition devised by Dr E. Evans (in photograph) to celebrate the 75th Anniversary of the founding of the Society.

Fig. 8.9. An historical moment caught at the Anniversary exhibition when it was set up in Cambridge in 1986: Dr T. Moore (lately Deputy Director of the Dunn Nutritional Institute, Cambridge) comes face to face with his father Professor B. Moore, F.R.S., the first Professor of Biochemistry in the U.K. and the founder of the *Biochemical Journal.*

The exhibition was presented at Liverpool, Cambridge, Dublin and London.

A History of the Society (this book) was commissioned by the main Committee and articles on the development of the Society were published in the *Biologist* [2] and *TIBS* [3]; the latter also carried an article on the proposed Biochemistry Gallery in The Science Museum, South Kensington [4].

To bring Biochemistry more directly to the notice of Schools an Essay Competition, supported by the *New Scientist*, was held. The topic was "Biochemistry and Society — Now and in the Future". A satisfactory but by no means overwhelming number of essays was received and the winner was Elizabeth Normand. Her essay was published in the *Bulletin* [5]. A poster prepared for the Society by ICI was distributed to Schools. At a lower level special commemorative stationery, beer mats and coffee mugs and T-shirts were available and a paper weight was presented to all contributors to the Society's 1986 Symposia, Colloquia and special lectures. Strictly speaking the T-shirts were not part of the 75th celebrations but were sold to promote the Biochemical Society Book Scheme to provide text-books for children in the Third World.

Apart from the statutory meeting in Scotland (Dundee in March) the three meetings in 1986 were held in centres historically closely connected with the Society: Liverpool (April), Cambridge (July) and University College London (December) (see Chapter 2). Additional money was made available to enable Groups organizing Symposia and Colloquia to invite more than one overseas participant to each meeting. A celebratory plaque was presented by the American Association of Biological Chemists at the Dundee meeting. At the Cambridge meeting Professor Yasutomi Nishizuka, on behalf of the Japanese Biochemical Society, presented the Society with a commemorative scroll (Fig. 8.10). Professor E. A. Dawes, Member of the Magic Circle, demonstrated his consummate conjuring ability.

The special anniversary dinner was held in December 1986 at UCL. 250 members were present and the occasion was graced by many representatives from overseas biochemical societies. The main guest speaker was Sir Hans Kornberg (Fig. 3.25). An anniversary address on behalf of the Jugoslav Biochemical Society was presented by Dr Elsa Reiner.

8.8 Education

Apart from the important business of organizing refresher courses PEC has devolved the organization of the general educational programmes of the Society to the Education Group. The Group's activities are discussed in detail in Chapter 5. Two important general developments have recently

Fig. 8.10 Professor Yasutomi Nishizuka, on behalf of the Japanese Biochemical Society, presenting a Commemorative scroll to the Society at the Meeting Dinner at Cambridge in July 1986. Professor H. M. Keir, Chairman of the Committee, is seen at the left.

Fig. 8.11. The 75th Anniversary dinner at UCL, December 1986.

taken place as the result of PEC initiatives. Firstly the Regional Groups set up some years ago to consider educational matters and which had gone into limbo have been revivified and reorganized. They are currently very active, not only on educational problems but on other topics referred to them by PEC. Secondly PEC has set up a joint working party with professional educationists to examine and advise on the presentation of Biochemistry in the A Level Biology syllabus. The decision to involve educationists was a wise one; indeed it was crucial if progress is to be made.

8.9 Animal Welfare

During the past two decades the general public has been made far more aware than previously of the use of experimental animals and has been encouraged to believe that this is not an activity compatible with a civilized society. There is no need to pursue this complicated problem further here but clearly the Biochemical Society is keen to see that the legitimate use of experimental animals should not be jeopardized as a result of acceptance of the distorted views of some pressure groups. Things came to a head when in 1979, two Bills of Parliament to regulate animal experimentation and to supersede the Cruelty to Animals Act 1976 were being proposed. The

Protection of Animals (Scientific Purposes) Bill was put forward by Mr Peter Fry under the auspices of the RSPCA and supported by the Animal Welfare groups while Lord Halsbury's Laboratory Animals Protection Bill was a product of the Research Defence Society (RDS).

The Biochemical Society, together with other similar academic organizations as well as the Pharmaceutical Industry, were asked to attend a meeting at the RDS to discuss a draft of Lord Halsbury's Bill. Professor P. N. Campbell, Director of the Courtauld Institute of Biochemistry, and Dr H. B. Waynforth, Head of the Animal Unit there, were asked by the Society to represent them. As a result of this meeting and the comments on it produced on behalf of the Society it was felt that the Society should have a specific member who could represent its interests in the field of laboratory animal welfare, use and legislation. Dr Waynforth was asked, and he accepted this position, which he holds at the time of writing. He sits on the Professional and Educational Committee. Dr Waynforth has kindly indicated the current developments:

> "Concurrently with the British Bills, the Council of Europe had produced a draft convention on the Protection of Animals Used for Experimental Purposes. The Society were asked to comment on the several drafts that became available and which were discussed by interested British organizations (including the Biochemical Society) at several meetings held at the Home Office.
>
> "The Halsbury and Fry Bills were allowed to proceed to various stages in Parliament but since the Government had declared its intention to produce its own Bill pursuant on the outcome of the Council of Europe Draft Convention, these did not proceed further. The Government's intentions were subsequently published in a White Paper "Scientific Procedures on Living Animals" in 1983 and comments from interested parties were requested. The Society's comments were forwarded and their usefulness acknowledged by the Home Office.
>
> "The Council of Europe draft convention set out standards for laboratory animal care which had to be adhered to by participating countries. However, it provided for individual national legislation which could ask for more stringent regulations and it was clear that the British Government would tread this path in several respects. The progress of the first White Paper and subsequently the more definitive second White Paper produced in 1985 was closely followed by the Society and comments were made to the Home Office at all the relevant stages. Unfortunately there was no feedback as to how influential these comments were, though it seemed clear that comments from the scientific organizations which, by all accounts, were fairly in accord, had far less impact than those of the British Veterinary Association, the politically motivated Committee for the Reform of Animal Experimentation and the Fund for the Replacement of Animals in Medical Experiments.
>
> "New regulations for animal experimentation became a reality on 20 May 1986 with the Queen's assent for the Animals

(Scientific Procedures) Act. This Act complements the provisions of the Council of Europe Convention which was finalized in 1985 and signed by the United Kingdom. The Act extends the Convention in several areas as expected.

"Although the new legislation is now *fait accompli*, matters of laboratory animal welfare and use are continually evolving and will concern animal-using members of the Society into the future. The unhealthy attitude of the militant 'animal rights' organizations makes it essential that the Biochemical Society, together with the other scientific organizations, keeps abreast of developments."

8.10 Postscript

As part of the 75th Anniversary celebrations the Society invited all surviving Officers of the Society to lunch late in 1985. Two buffet lunches were arranged and the afternoons were spent in informal discussion. The occasions were enjoyable and delightful; 'old timers' were delighted to see each other and their younger successors and the author of this History gained much valuable 'copy' from the discussions which were recorded and are now stored as part of the Society's archives. The photographer was also busy and a selection of his pictures are collected as Plates in this volume (pages 93–96). These two gatherings were so successful on all levels that one strongly commends their continuation to future Officers of the Society.

References

1. Baum, H. (1986) Higher education — what has gone wrong? *J. Roy. Soc. Med.*, **79**, 315–316.
2. Goodwin, T. W. (1986) The Biochemical Society 1911–1986. *Biologist*, **33**, 83–86.
3. Goodwin, T. W. (1986) The 75th Anniversary of the Biochemical Society. *TIBS*, **11**, 340–343.
4. Kamminga, H. (1986) Biochemistry on display. *TIBS*, **11**, 343–344.
5. Normand, E. (1986) Biochemistry and Society — now and in the future. *Biochem. Soc. Bull.*, **8**, no. 3, pp. 11–13.

Index

Abbreviations used in the Index are those identified at the beginning of the book (page iv). Numbers in **bold** after entries denote pages on which relevant illustrations (Figures) occur. Positions of Tables are given by page numbers followed by the Table number, after the letter T, in parentheses, e.g. 47(T 3.5).

Aberystwyth (University of Wales), emergence of Biochemistry in Wales   5
Abstracts, BS Meetings papers
   published in *Chemistry and Industry* (1929–1941)   35
   published in *BJ*   35, 81
Administration
   acquisition of headquarters   39
   full-time staff Secretary appointed   41
   staff organization   **44**
   Warwick Court Committee Room **43**   42
Administrative Secretary *see* Executive Secretary
Advisory Committee for Publications *see* Publications Board
Agenda Papers
   BJ subscriptions and membership   81
   introduction   61
Agricultural chemistry   3, 5
Aldridge, W. N.   **126**
   Chairman, *BJ* Editorial Board (1965–1969)   126
   first BS Poster Session   56
American Association of Biological Chemists, 75th celebratory plaque presented to BS   169
Animals, experimental, welfare of   170–172
Animals (Scientific Procedures) Act   171
Anniversary celebrations, BS
   50th Anniversary   51
   500th Meeting   52, 53
   75th Anniversary **93–96, 168, 170**   52, 75, 168, 169, 172
Ansell, G. B.   **141**
   Hon. Archivist, BS   75
   Neurochemical Club   106
   Publications Secretary   141
Archives, BS, establishment   74, 75
Armstrong, H. E.   **17**
   Biochemical Club Committee member   15
   inaugural meeting, Biochemical Club   14
   resignation from Biochemical Club   17
Arnstein, H. R. V.   **95**
   BS Hon. Secretary   47(T 3.5), 59, 60
   BS Meetings Secretary   154
   Chairman of PESC   165
   foundation of Medal awards   66, 67
   Secretary-General of FEBS   154
Articles of Association, BS
   incorporation of BS   45
   modification after Extraordinary General Meeting   82
   run-on costs of journals   92
Awards Committees, BS   69

Bacharach, A. L., BS member's anecdote   60
Bachelard, H., Neurochemical Group, BS   105, 106

Baddiley, James, first Chairman of Carbohydrate Group, BS   110
Baker, J. L., BS Trustee   29, 46(T 3.3)
Bangor (University of Wales), emergence of Biochemistry in Wales   5
Barger, G., BS Chairman   27(T 2.5)
Barker, G. R.   **44**
   adviser, BS Mobile Exhibition   168
   Hon. Archivist, BS   75
   Joint Nucleotide Group, BS   108
Baum, H. **158**   BS Hon. Public Relations Officer   158, 165
Bayliss, W. M. (Sir William Bayliss)   **17**
   Biochemical Club Committee member   15(T 2.1)
   BS Chairman   27(T 2.5)
   Chairman at first Biochemical Club A.G.M. (1912)   17
   Editor of *BJ* (1912–1924)   23
BDH Chemicals Ltd. Award in Analytical Biochemistry   67
Beechey, R. B. **138**, Managing Editor, *BS Transactions*   137
Beedle, A. S., staff Deputy Editorial Manager   127
Biochemical Club
   acquisition of *BJ*   18–23
   change of title to Society   23, 26
   first A.G.M. (1912)   17, 17(T 2.3)
   founding Committee   15(T 2.1)
   inaugural meeting   13, 14
   men only membership rule   14
   venues of meetings (1911, 1912)   16(T 2.2)
   visit to Rothamsted Experimental Station (1911) **16**   15–17
Biochemical Immunology Group, BS   107
Biochemical Journal (BJ)
   1945–1965   116–125
   1965–1986   125–132
   acquisition by Biochemical Club (1912)   18–23
   annual number of pages   122(T 6.2)
   annual number of papers: published 1906–1942 **118**; submitted 1955–1985 **133**
   appeal for funds   119
   BS finances   78–88
   BS member's subscription   80, 81
   C. U. P. as publisher   20, 23, 28, 124
   change of printer   123–125
   costing   89–92
   Editorial Advisory Panel   120
   Editorial Committee   119, 126
   editorial procedures   **132**
   editors, first   23
   editors' recollections   116–118, 121, 122
   finances: early, 27, 28; post-second World War, 48, 78, 79, 115
   first Volume, Contents   19(T 2.4)

INDEX   173

Biochemical Journal — contd.
    foundation of, by Benjamin Moore at Liverpool
        (1906)   8, 10, 18
    frequency of publication   120
    future   145
    Harden Bequest   70
    international origins of papers   128, 129(T 6.1)
    Letters   127, 131
    Liverpool University Press as publisher   18
    profitability   90, 91, 119
    promotion   128, 129, 165, 167
    publication of BS Abstracts   35, 137
    publication times   **128**
    publicity   128, 129, 165, 167
    Rapid Papers **134**   130
    Reviews   127, 131
    revised pricing policy   90, 91
    run-on cost   81–83, 92
    sales   92
    sectionalization   126, 129
    Short Communications   130
    subscribers **90**   91, 92(T 4.4)
    subscription rates   28, 90, 91
    see also Editorial Board, *BJ*; Editorial Office
Biochemical Society Bulletin
    development from Agenda Papers   61
    typical Contents   143, 144(T 6.2)
*The Biochemical Society, its History and Activities
    1911–1969*, publication 52
Biochemical Society Symposia
    500th BS Meeting Symposium   52
    launch   54, 143
    organizers   55(T 3.8)
    sales **143**   143
Biochemical Society Transactions
    BS finances   87
    BS member's subscription rate   82, 85
    evolving policy   138, 139
    FEBS Meeting special issue   155
    foundation   115, 136–139
    promotion   165
    publication of BS Proceedings   55
    subscribers **91**   91
Biochemische Zeitschrift, founded 1906   18
Biochemistry
    BS textbook   144
    Courses in   160
    early development: in Germany, 1, 3, 13, 18; in U.S.A.,
        2, 18; in U.K., 3–11
    employment in   158–161
    integral position in biology   34, 170
    modern specialization   141
    post-Second World War development   37
    public reaction   165
Biochemistry Gallery, Science Museum   166, 169
Bioenergetics Group, BS   111
Biological Council, formation   50, 52
Bioscience Reports
    Board of Editors   140
    foundation   116, 140
    transfer of copyright   140
Biotechnology
    British Coordinating Committee for   110
    careers in   160
Blackman, V. H., BS Chairman   27(T 2.5)
Blaschko, Hermann, BS member, Neurochemical Group,
    BS   106

Boehringer-Mannheim Travelling Fellowship   68
Book Distribution Depot (Colchester)
    BS distribution service for learned societies   88
    purchase of premises   42, 88
Book Scheme, BS   169
Boyce, Sir Rupert, George Holt Professor of Pathology
    (Liverpool)   6
Boyd, G. **163**
    Chairman of PESC   163
    Refresher Courses   163, 164(T 8.3)
British Biophysical Society   50(T 3.6), 153
British Coordinating Committee for Biotechnology   110
British National Committee for Biochemistry
    Chairmen (1956–1982)   152(T 7.1)
    formation   151, 152
    travel funds   63
Brown, A. J., Biochemical Club Committee
    member   15(T 2.1)
Burdon, R. H. **93, 96**
    BS Hon. Meetings Secretary   59
    BS Hon. Secretary   47(T 3.5)
Bushill, J. H. **77**
    BS Hon. Treasurer (1944–1952)   45, 77, 78, 120
    Convenor of BS Trustees   45, 46(T 3.3)
Butenandt, A. F. J., Nobel Prizewinner   34(T 2.7)
Bywaters, H. W., original BS member   **32**

Cambridge, University of
    Dunn Biochemistry Institute (1924)   10
    Physiological Chemistry taught (1895)   3
    third U.K. Chair (Sir William Dunn Chair) of Bio-
        chemistry   10
Cambridge University Press
    approached as *BJ* publisher (1911)   20
    BJ printers (1985)   125
    commission on *BJ* sales   119, 121–124
    contract for *BJ*   23, 28
Campbell, P. N. **93**
    BS Hon. Secretary   41, 47(T 3.5), 153
    Editor, *Essays in Biochemistry*   141, 142
    Laboratory Animals Protection Bill   171
    organization of European Meetings   154
Carbohydrate Group, BS   109
Careers Adviser, BS   158
Careers Conferences   161, 162
Careers in Biochemistry, Careers for Biochemists   159
Cathcart, E. P., Professor (Gardiner Chair) of Physiological
    Chemistry (Glasgow 1919)   4
'Cells, Molecules and Life', 75th BS Anniversary Exhibi-
    tion   75, 166
Chairmen, BS
    1911–1944   27(T 2.5)
    1945–1986   47(T 3.4)
    founder Chairman   26
Channon, H. J. **24**
    BS Chairman   23, 27(T 2.5)
    Discussion Meetings   34
    Professor (Johnston Chair) of Biochemistry (Liverpool
        1932)   8
Chemical Society, The
    distribution of *BJ* for BS   124
    joint Discussion Meeting with BS   34
    Library subscription, BS   73, 74
Chemistry and Industry, publications of BS Abstracts
    (1924–1941)   35
Chibnall, A. C. **27, 72, 93**
    BS Chairman   47(T 3.4)

Chibnall, A. C. — *contd.*
   BS Hon. Member   71(T 3.16)
   BS Hon. Secretary   27, 27(T 2.6)
   BS Trustee   46(T 3.3)
   President of International Congress of Biochemistry   148
   Professor (Sir William Dunn Chair) of Biochemistry (Cambridge)   27
Chick, Harriet (Dame Harriet Chick), elected BS member (1913)   15
Chinese Biochemical Society **156**, delegation to BS Meeting (1982)   156
CIBA Research Laboratories, financing of CIBA Medal **66**   66
Clark, F. **121**, first staff Secretary to *BJ* Editorial Board   121, 126
Clinical Science
   Committee of Management   134
   Editorial Board   135
   joint publication BS/Medical Research Society   87, 132
   Supplements   136
   Trustees   134, 135
Close, J. K., original BS Member   **32**
Colchester Depot *see* Book Distribution Depot
Colloquia, BS.   56
Colworth Medal **66**   66
Committee, Biochemical Club (1911-1912)   15(T 2.1)
Committee (Main Committee), BS
   Articles of Association   45, 46
   conflict with *BJ* Editorial Board   124, 125, 127
   future organization   49
   policies   49
   post-Second World War planning   48, 49
Companies Act, BS Incorporation under   45
Conferences, BS   162
Conway, E. J., Professor of Biochemistry and Pharmacology (National University of Ireland)   5
Cowell, S. J., Associate Editor *BJ*   116
Cramer, W., original BS member   **32**
Crook, E. M., Symposium Organizer, BS   55(T 3.8)
Cuthbertson, D. P. (Sir David Cuthbertson) **95**, BS Chairman   27(T 2.5)
Cuthbertson, W. J. F., **80**, BS Hon. Treasurer (1962-1972)   46, 80, 81, 154

Dale, G. **94**, staff Accountant   89
Dale, H. H. (Sir Henry Dale)
   50th BS Anniversary   52
   Biochemical Club Committee member   15(T 2.1)
   BS Hon. Member   71(T 3.16)
   Nobel Prizewinner   34(T 2.7)
Dalgliesh, C. E. **95**, BS Hon. Secretary   47(T 3.5)
Datta, S. P. **154**, Treasurer FEBS   154
Davidson, J. N.
   50th BS Anniversary   52
   BS Chairman   47(T 3.4), 124
   BS Hon. Secretary   47(T 3.5)
   founding of IUB   151, 152
   Professor (Gardiner Chair) of Physiological Chemistry (later Biochemistry) (Glasgow 1948)   4
Davison, A. N. **93**, BS Hon. Secretary   47(T 3.5), 106, 155
Davy, Humphrey, 'Agricultural Chemistry' (1802)   3
Dawson, R. M. C. **96**, Chairman of Publications Board, BS   6, 140
Day, Shirley **88**, staff Business Manager   88

de Duve, C. R. M. J., BS Hon. Member   71(T 3.16)
Developments initiated, BS   50(T 3.6)
Dickens, F. **93**
   Associate *BJ* Editor   116
   BS Chairman   47(T 3.4)
   BS Hon. Member   71(T 3.16)
   Editor, *Essays in Biochemistry*   142
Dingle, J. **127**, Chairman, *BJ* Editorial Board (1975-1982)   127
Discussion Meetings, BS
   inauguration   32
   publication as *BS Symposia*   54
Distribution service, BS   88
Dixon, M., BS Hon. Member   71(T 3.16)
Dodds, E. C. (Sir Charles Dodds) **150**
   BS Chairman   27(T 2.5), 47(T 3.4)
   BS delegate to IUPAC Congress (1951)   150
   BS Hon. Member   71(T 3.16)
Dodgson, K. S. **158**
   BS Chairman   47(T 3.4)
   BS Hon. Secretary   47(T 3.5)
   careers booklet   158
Dorée, C., original BS member   **32**
Drummond, J. C. (Sir Jack Drummond)
   BS Chairman   27(T 2.5)
   BS Hon. Secretary   27(T 2.6)
   BS Trustee   46(T 3.3)
   special *BJ* Sub-committee member   120
Dudley, H.W. **23**
   Assistant Editor *BJ* (1924-1930)   23
   BS Chairman   27(T 2.5)
   BS Hon. Secretary   27(T 2.6)
   BS Trustee   24, 46(T 3.3)
Dunn Biochemistry Institute (Cambridge 1924)   10
Dyer, B. **32**, BS Chairman   27(T 2.5)

Edinburgh University, Chair of Chemistry Related to Medicine (1929)   4
Editorial Advisers Panel, *BJ*   120, 127
Editorial Board, *BJ*
   appointment of staff Editorial Secretary   121
   formation   49, 120
   growth in numbers   126
   threatened resignation   124
Editorial Committee, *BJ*
   BJ finances   119
   formation   126
Editorial Office
   first staff Editorial Secretary   121
   locations: 1946-1952, Postgraduate Medical School, 40; 1952-1955, National Institute for Medical Research, 40, 121; 1955-1960, Lister Institute of Preventive Medicine, 40; 1960-1961, no. 133-135 Oxford Street 40; 1961-1966, Medical Research Council (Park Crescent), 41; since 1966, no. 7 Warwick Court, **41**, 41
   reorganization   127, 132
Education Group, BS   112
Educational activities
   Education Group, BS   112
   educational programmes   169
   Refresher Courses   163, 164(T 8.3), 169
Elliott, D. F. **80**
   BS Hon. Treasurer (1972-1982)   42, 83-89
   funding of CIBA Medal and Prize   66
Ellis, G. W., 50th BS Anniversary   52
Emeritus Members, BS   39

Employment in Biochemistry
    career   158–161
    register   162
    surveys   160, 161(T 8.2)
Essays in Biochemistry, foundation   115, 141
Essays in Medical Biochemistry, foundation and discontinuation   116, 142
European Federation of Biotechnology, BS representatives   110, 156
European Society for Neurochemistry, formation   106
Evans, A.(Tony) G. J. **94**, staff Editorial Manager   43, 127
Evans, Charles Lovatt (Sir Charles Lovatt Evans)   **32**
    50th BS Anniversary   52
    BS Chairman   27(T 2.5)
    first paid-up member, Biochemical Club   15
Evans, Elizabeth **168**, design of BS Stands at international Meetings   166–168
Executive Secretaries (staff)   41, 43, 73
Experimental animals, welfare of   170–172
Eyre, J. V. **32**, BS Chairman   27(T 2.5)

Fabre, R., French Biochemical Society   55
Faraday Society, joint Meeting with BS   34
Fearon, W. R., Professor of Biochemistry (Trinity College Dublin 1934)   5
Federation of European Biochemical Societies (FEBS)
    BS Stands at Meetings   166
    Congress Funds   63
    inauguration   130, 153
Fellowships, BS   68, 69
Finance Board, BS, formation   49
Finances, BS
    1944–1986   77–92
    acquisition of headquarters building   81
    attitudes to surpluses   78, 79
    BJ as source of income   28, 48
    currency exchange rates   92
    expenditure votes (1970 and 1979) 83(T 4.1)
    incorporation   29
    inflation effects   82, 83
    investment policies   45, 79, 80, 87, 89
    legal status   45, 46
    Meetings revenue and costs   86(T 4.2)
    non-profit lobby   78
    revenue summaries (1985)   89(T 4.3)
    subscription income   79
    Trustees   29
Findlay, G. M., BS Chairman   27(T 2.5)
Florkin, Marcel   151
Foster, M., Professor of Physiology (Cambridge 1880s)   9

Gaddum, Sir John, BS Hon. Member   71(T 3.16)
Gardner, J. A. **13, 32**
    Biochemical Club Committee member   15(T 2.1)
    BS Hon. Treasurer (1913–1944)   26, 27, 77
    BS Trustee   29, 46(T 3.3)
    formation of Biochemical Club   13
    seeking BS journal publisher   20
Gardner, T. E. (brother of J. A. Gardner), assistance to BS Treasurer   26, 28
Garland, P. B., Symposium Organizer   55(T 3.8)
Garrod, A. E., Biochemical Club Committee member   14
Germany, early development of Biochemistry   1, 2
Gesellschaft für Physiologische Chemie (1947), renamed Gesellschaft für Biologische Chemie (1964)   2

Glasgow University, Physiological Chemistry, lecturership (1905) and Chair of Physiological Chemistry (1919)   4
Golding, J., original BS Member   **32**
Goodwin, T. W. **93**
    BS Chairman   47(T 3.4)
    BS Hon. Member   71(T 3.16)
    Morton Lecturer   66
    Symposium Organizer   55(T 3.8)
Graduates in Biochemistry, BS Survey   160
Grant, J. K., Symposium Organizer   55(T 3.8), 99
Greville, G. D., Editor, *Essays in Biochemistry*   141
Grieve, J., lecturership bequest, Physiological Chemistry (Glasgow University 1905)   4
Griffiths, J. Jones, lecturer in Agricultural Chemistry (Aberystwyth 1906)   5
Groups, BS
    guide lines   100, 101
    introduction and importance   48, 56, 97–99
    Meetings   56–57
    titles   102(T 5.1)
Grünbaum, A. S., lecturer in Physiological Chemistry (Oxford 1898)   4
Gutfreund, H., proposer of Molecular Enzymology Group, BS   97

Haas, P., BS Chairman   27(T 2.5)
Haldane, J. B. S., BS member's anecdote   32
Haldane, J. S., lecturer in Physiological Chemistry (Oxford)   3
Halliburton, W. D. **3**
    Biochemical Club Committee member   15(T 2.1)
    first BS Hon. Member   3, 31
    formation of Biochemical Club   14
    Professor of Physiology (King's College London)   3
Happold, F. C. **154**, BS Chairman   47(T 3.4), 154
Harden, A. (Sir Arthur Harden)   **10, 32**
    at inaugural Meeting of Biochemical Club   14
    Biochemical Club Committee member   15(T 2.1)
    BS Chairman   27(T 2.5)
    BS Hon. Member   25, 30, 32
    BS presentation **24**, 23
    BS Trustee   29, 46(T 3.3)
    career   24–26
    Editor, *BJ* (1913–1937)   23, 27, 116
    legacy (Harden Conference)   26, 70
    Nobel Prizewinner   25, 32, 34(T 2.7)
    Professor of Biochemistry (Lister Institute of Preventive Medicine 1912)   10, 25
Harden Conferences   26, 70
Harden Medal   65
Harington, C. R. (Sir Charles Harington) **23**
    50th BS Anniversary   52
    500th BS Meeting   **72**
    assistant to Editor, *BJ* (1930–1937)   23, 27
    BS Chairman   47(T 3.4)
    BS Hon. Member   71(T 3.16)
    BS Trustee (Chairman)   45, 46(T 3.3)
    Director, National Institute for Medical Research   116
    Editor, *BJ* (1930–1942)   116
    formation of IUB   150
    news report of Communication   62
Harrington, M. G., Irish Area Section, BS   104
Hartley, B. H., BDH Award Lecturer   67
Hartley, P.
    BS Chairman   27(T 2.5)
    BS Hon. Secretary   27(T 2.6)

Hartree, E. F., BS Hon. Member   71(T 3.16)
Haslewood, G. A. D. **52, 93**   47(T 3.4), 52
Hatschek, E., BS Chairman   27(T 2.5)
Haworth, Sir Norman, Nobel Prizewinner   34(T 2.7)
Heads of Departments Conferences   162, 163
Hele, T. S., BS Chairman   27(T 2.5)
Hemming, F. W. **160,** Chairman PESC   160
Henry, T. A. **32,** BS Chairman   27(T 2.5)
Henton, A. I. P. **41**, staff Executive Secretary   41, 46, 73
Herriott, Doris **94**, staff Meetings Officer   43, 60, 167
Hill, A. V., Nobel Prizewinner (1923)   34(T 2.7)
Hofmeister's Beitrage, founded 1901   18
Honorary Members, BS
  1944–1986   71(T 3.16)
  at 500th BS Anniversary meeting   **72**
  first elected   3
  plaque   **71**
  rules for election   71
Honorary Officers, BS
  Chairmen 1914–1944   27(T 2.5)
  first appointed   26
  Secretaries: 1911–1945,   27(T 2.6); 1945–1986, 47(T 3.5)
  terms of appointment   28
  Treasurers: 1913–1944,   26, 27, 77; 1944–1952,   45, 77, 78; 1952–1986, 78–89
Hopkins, F. Gowland (Sir F. Gowland Hopkins) **4, 32**
  Biochemical Club Committee member   15(T 2.1), 20
  BS Hon. Member   32
  first BS Chairman   26, 27(T 2.5)
  laboratory teaching   29
  Lecturer in Chemical Physiology (Cambridge 1895)   3, 8, 9
  Nobel Prizewinner (Medicine)   30, 34(T 2.7)
  Professor (William Dunn Chair) of Biochemistry (Cambridge)   10
  regard for Harden's *BJ* editorship   23
Hopkins Memorial Lecture   **64**   64
Hoppe-Seyler's Zeitschrift für physiologische Chemie, founded 1877   2, 18
Hormone Group, BS   108
Hurtley, W. H., original BS member   **32**

Incorporation   45, 80
Industrial Biochemistry and Biotechnology Group, BS   110
Industry, liaison with   160
Institute of Biology, formation   51
Institute of Brewing, joint Meeting with BS (1923)   33
International Congresses   147, 148, 150
International Meetings, presence of BS **167**   166, 167
International Secretaries, BS   47, 154, 155
International Society of Neurochemistry, formation   55
International Society of Scientific Unions (ICSU), foundation of IUB   151, 152
International Union of Biochemistry (IUB)
  BS Stands at Meetings **167**   166
  Congress Funds   62
  foundation   149–152
International Union of Pure and Applied Chemistry (IUPAC), foundation of IUB   147, 149–152
Introducing Biochemistry   144
Investment policies   79, 80
Irish Area Section, BS
  formation   101, 104
  Irish Group, oral presentations   57
  Irish Lecture Tour   104

Irish Universities *see* National University of Ireland; Queens University, Northern Ireland; Trinity College Dublin

James, A. T.
  foundation of Colworth Medal   66
  Lipid Group, BS   107
Japanese Biochemical Society
  75th BS Anniversary celebration **169**   169
  joint Colloquium with BS   155
Jenner Institute *see* Lister Institute of Preventive Medicine
Johnston Chair of Biochemistry (Liverpool 1902)   6
Jones, D. S. **159**, Careers Adviser BS   158, 160
Jones, G. D. **94**, staff Executive Secretary   43
Journal of Biological Chemistry, founded 1905   18
Journal of the Chemical Society, biochemical papers   18
Journals, BS, *see under individual titles*
Jubilee Lecture   64
Jugoslav Biochemical Society, presentation to BS   169
Junior Travelling Fellowships, BS   67

Kamminga, H., Science Museum BS Exhibition   166
Karrer, P., Nobel Prizewinner (1937)   34(T 2.7)
Kay, H. D.
  BS Hon. Secretary   27(T 2.6)
  BS Trustee   29, 46(T 3.3)
Kay, J., Symposium Organizer   55(T 3.8)
Keeble, F. **20**, Biochemical Club Committee member   15(T 2.1), 20, 21
Keilin, David, Memorial Lecture   65
Keir, H. M. **93, 169**
  BS Chairman   47(T 3.4)
  Careers Adviser, BS   160
Kennaway, E. L., original BS member   **32**
Killip, J. D., staff Secretary to *BJ* Editorial Board   127
King, E. J. **40**
  BS Chairman   47(T 3.4)
  Chairman, *BJ* Editorial Board (1946–1952)   40, 120
King's College London, Chair of Physiology   3
Klyne, W., Editor *BJ*   120
Knight, B. C. J. G., special *BJ* Sub-committee member   120
Kornberg, Sir Hans   **75**
  guest speaker at 75th BS Anniversary Dinner   169
  lecturer at Science Museum BS Exhibition   75
Kornberg, Lady, at Science Museum BS Exhibition   75
Krebs, H. (Sir Hans Krebs) **52, 72**
  80th Birthday Colloquium   155
  at 500th BS Meeting   72
  Boehringer–Mannheim Travelling Fellowship in honour of   68
  BS Hon. Member   71(T 3.16)
  Memorial Scholarship   69
  Sub-committee report   143
  Symposium in tribute   55
  third Hopkins Memorial Lecture   51
Krebs Memorial Scholarship   69
Krogh, S. A. S., Nobel Prizewinner (1920)   30, 34(T 2.7)

Laboratory animals, welfare   171
Langley, J., Professor of Physiology (Cambridge)   9, 10
Lea, Sheridan, Lecturer in Chemical Physiology (Cambridge 1895)   3, 9
Lectures and Memorial Lecturers   64, 65
Letters, R., Irish Area Section, BS   105
Lipid Group, BS   107

Lister Institute of Preventive Medicine (formerly Jenner Institute)
    Harden, acting Director (1914–1918)   25
    location of *BJ* Editorial Office (1955–1960)   40
    second U.K. Chair of Biochemistry (1912)   10
    venue for BS Discussion Meeting   32
Liverpool, University of, first U.K. Professorship (Johnston Chair) of Biochemistry (1902)   4, 6–8
Liverpool University Press, first publishers of *BJ*   18
Lloyd, J. B., BS Hon. Secretary   47(T 3.5)
Loewi, O., Nobel Prizewinner (1936)   34(T 2.7)
logos, BS   76
London, University of, second U.K. Chair of Biochemistry (Lister Institute 1912)   10
London University Press, approached to publish *BJ* (1911)   20
Lucy, J. A. **131**, Deputy Chairman, *BJ* Editorial Board   131

Macleod, J. J. R., Nobel Prizewinner (1923)   34(T 2.7)
MacFarlane, M. G., publication of *Essays*   141
McIlwain, Henry, Symposium Organizer   102
Main Committee *see* Committee, BS
Malcolm, A. D. B., BS Hon. Secretary   47(T 3.5)
Mann, Sydney A. **32**, Founder Member, BS   105
Martin, Sir Charles **32**
    anecdote about J. B. S. Haldane   32
    BS Hon. Member   71
    original BS Member   **32**
Mathias, A. P. **93**
    BS Hon. Secretary   47(T 3.5)
    BS International Secretary   47, 155
    BS Meetings Secretary   155
Mears, H., colleague of J. H. Bushill   78
Medals and medallists   65–67, 107
Medical Research Council, location of *BJ* Editorial Office (Park Crescent) (1961–1966)   40
Medical Research Society
    joint publishers with BS of *Clinical Science*   87, 133
    meetings Communications   136
Meetings, BS
    21st Birthday (17 November 1933)   29
    100th (13 March 1926)   29, 31
    500th   52, 53
    Agenda Papers   61, 81
    Anniversary   51–53
    attendance fees   86
    Discussion   32, 54, 55
    frequency   54
    international   62
    joint   33
    Meetings Board   49
    ordinary   54–61
    publication of Abstracts   35, 81
    Regional Careers Conferences   161
    Second World War programme   54, 56, 57
    typical schedule   58(T 3.9)
Mellanby, E., original BS member   **32**
Membership, BS
    apathy to Society affairs   98
    categories **38**   38, 80
    cost of services   85
    early days   29, 30
    Emeritus   39
    Honorary   30–32
    international   39(T 3.2)
    Nobel Prizewinners   34(T 2.7), 72(T 3.17)

Membership, BS — *contd.*
    original   **32**
    subscriptions   79, 82, 119
Membrane Group, BS   112
Meredith, W. M., valuation of *BJ* (1912)   21, 22
Miers, Principal, Manchester University valuer of *BJ* (1912)   21
Mills, A. K., Guinness Research Laboratories   105
Milroy, J. A., Professor of Biochemistry (Queen's University, Northern Ireland 1924)   5
Milroy, T., Lecturer in Advanced Physiology and Physiological Chemistry (Edinburgh)   4
Mobile Exhibition, BS **168**   166
Molecular Enzymology Group, BS   97, 99, 102(T 5.1)
Moore, Benjamin **7**
    Biochemical Club Committee member   15(T 2.1)
    elected F.R.S. (1912)   8
    founder of *BJ*   8, 18
    introduced first U.K. Honours School of Biochemistry (Liverpool 1910)   8
    Lecturer in Physiology (Charing Cross Medical School)   7
    Physiology Department (UCL)   7
    Professor (Johnston Chair) of Biochemistry (Liverpool 1902)   6, 7
Moore, T. (son of Benjamin Moore), quoted on *BJ* royalties   18
Morgan, Walter T. J. **72, 96**
    at 500th BS Meeting   72
    BS Chairman   47(T 3.4)
    BS Hon. Member   71(T 2.6)
    BS Hon. Secretary   27(T 2.6)
    Carbohydrate Group, BS   109
    formation of Biological Council   51
    Meetings programme   54
    special *BJ* Sub-committee member   120
Morton, R. A. **52, 72**
    50th BS Anniversary   52
    500th BS Meeting   52
    author of BS *History* (1911–1969)   52, 147
    BS Chairman   47(T 3.4)
    BS Hon. Member   71(T 3.16)
    Johnston Professor of Biochemistry (Liverpool)   65
    Memorial Lecture   65
Morton Lecture   65
Murray, Margaret M., BS Chairman   47(T 3.4)

National Institute for Medical Research
    Benjamin Moore (1914)   8
    Charles Harington, Director   116
    location *BJ* Editorial Office (1952–1955)   40
National University of Ireland
    BS Irish Area Section Meetings   104
    Department of Biochemistry and Pharmacology (1934)   5
Needham, Dorothy M., BS Hon. Member   71(T 3.16)
Neuberger, A. **95**
    BS Chairman   47(T 3.4)
    BS Hon. Member   71(T 3.16)
    Chairman, *BJ* Editorial Board (1952–1955)   40, 121
Neurochemical Group, BS   101, 105–107
Nishizuka, Yasutomi **169**
Nobel Laureates, BS
    1911–1945   34(T 2.7)
    1945–1986   72(T 3.17)
    special celebratory Dinner (3 February 1930) **33**,   30

Nowell, P. T., Pharmacological Biochemistry Group   103, 104
Nucleic Acid and Molecular Biology Group, BS   109
Nucleotide and Nucleic Acid Group, BS   108
Nutrition Society, joint Meeting with BS (1944)   33

Ogston, A. G. **94**, Chairman, *BJ* Editorial Board (1955-1959)   121
Osler, Sir William, donor for BS purchase of *BJ*   22
Oxford, University of
    Physiological Chemistry (1897)   3
    Wayneflete Chair of Physiology (1914)   4
Oxford University Press, approached to publish *BJ* (1911)   20

Page, H. J. **32**, BS Chairman   27(T 2.5)
Paine, S. G., original BS Member   **32**
Parke, D. V., Pharmacological Biochemistry Group, BS   103
Pasternak, C. A., Managing Editor, *Bioscience Reports*   140
Pasteur Medal, presentation to BS   55
Pathological Society, joint Meeting with BS (1944)   33
Peptide and Protein Group, BS   111
Perham, R., Peptide and Protein Group, BS   111
Perry, S. V. **96**
    BS Chairman   47(T 3.4)
    BS Hon. Member   71(T 3.16)
Peters, R. A. (Sir Rudolph Peters) **51**
    BS Chairman   27(T 2.5), 47(T 3.4)
    BS Hon. Member   71(T 3.16), 105, 106
    BS Trustee   46(T 3.3)
    Chairman, British National Committee for Biochemistry   152
    first Hopkins Memorial lecturer   64
    formation of Biological Council   51
Pharmacological Biochemistry Group, BS   99, 102-104, 102(T 5.1)
Phelps, C. E., Symposium Organizer   55(T 3.8)
Philosophical Transactions of the Royal Society   3
Physiological chemistry, German origins of   1, 3
Physiological Society
    biochemical papers at Meetings   17
    International Congress of Physiology   147
    joint Meetings with BS   33, 34
    rules used as model by Biochemical Club   14
Pirie, N. W.
    anecdote about J. B. S. Haldane   32
    special *BJ* Sub-committee member   120
Plimmer, R. H. A. **3, 32**
    author of BS *History* (1911-1949)   14
    Biochemical Club formation   13-15
    BS Chairman   27(T 2.5)
    BS Hon. Member   32
    BS Hon. Secretary   26, 27(T 2.6)
    Founder Member, BS   3
    Hon. Secretary/Treasurer Biochemical Club   15
    investment policy, BS   79
    original BS Trustee   29, 46(T 3.3)
    seeking BS journal publisher   20
Pogson, C. I. **128**, Chairman, *BJ* Editorial Board (1982-1987)   128
Porter, Helen **46**, BS Chairman   46, 47(T 3.4), 98
Porter, R. R. **98**
    Biochemical Immunology Group, BS   107

Porter, R. R. — *contd.*
    BS Chairman   47(T 3.4)
    BS Hon. Member   71(T 3.16), 98
Poster Sessions, inception **57**   56
Powell, Gillian **159**, Careers Adviser, BS   159
Proceedings, BS
    early Abstracts   35, 81
    separation from *BJ*   82
Professional and Educational (Sub-)Committee (PESC)
    careers and employment   158-161
    Chairmen   158(T 8.1)
    formation   48, 49, 157, 158
    Royal Charter petition considered   73
Promotions Organizer, BS   76, 164, 165
Properties owned, BS **41**   41, 42
Protein Group, BS/Chemical Society Joint Group   111
Pryde, J., Professor of Biochemistry (Welsh National School of Medicine 1956)   5
Publications Board, BS
    book publishing   143
    founded as Advisory Committee for Publications   48, 49, 125
    proposals for *BS Transactions*   137
Publications Secretaries   140, 141
Public relations, BS   158, 166, 169
Publishing activities
    BS Bulletin   143
    BS histories   14, 52, 169
    BS logo   76, 144
    careers booklets   159
    collaboration with commercial publishers   140, 144
    Essays   141, 142
    future   145
    journals   116-141
    rapid publication journal   130, 140
    special publications   143, 144, 159
    Symposia   116, 142, 143
Pyman, F. L., BS Chairman   27(T 2.5)

Quastel, J. H. **105**, BS Hon. Member   71(T 3.16), 105
Queen's University Northern Ireland
    autonomous Department of Biochemistry   5
    BS Irish Area Section Meetings   104
    Professorship (J. C. White Chair) of Biochemistry   5

Raistrick, H.
    BS Chairman   47(T 3.4)
    BS Meetings Secretary   27
    BS Trustee   46(T 3.3)
Ramsden, W. **8, 32**
    Biochemical Club Committee member   15(T 2.1)
    Lecturer in Physiological Chemistry (Oxford 1897)   3
    Professor (Johnston Chair) of Biochemistry (Liverpool)   8
    seeking BS journal publisher   20
Rapid Papers, *BJ*   130, 140
Refresher Courses   163, 164(T 8.3), 169
Regional Careers Conferences   161, 162
Regulation in Metabolism Group, BS   112
Research and Information Officer, staff   165
Reviews in *BJ*   131
Richter, Derek, Neurochemical Group, BS   106
Robinson, D. **93**, BS Hon. Secretary   47(T 3.5)
Robinson, F. A. **78**
    BS Chairman   47(T 3.4)
    BS Hon. Treasurer   78-80, 123

Robinson, Sir Robert, BS Hon. Member   71(T 3.16)
Robison, R.
  BS Chairman   27(T 2.5)
  BS Hon. Secretary   27(T 2.6)
Robson, W.
  BS Chairman   47(T 3.5)
  BS Hon. Secretary   27(T 2.6)
Rogers, H. J. **95**, Chairman, *BJ* Editorial Board (1964-1969)   125
Rothamsted Experimental Station, Biochemical Club visit   15-17
Roughton, F. J. W., assistant to Editor, *BJ*   23, 116
Royal Charter, BS consideration of   73
Royal Commission on Medical Education, BS memorandum   112
Royal Society, The
  donations to *BJ* publication   119
  formation of IUB   151
  formation of Society for the Promotion of Animal Chemistry   3
  travel funds   62
Russell, E. J., at inaugural Meeting of Biochemical Club and Committee member   14, 15(T 2.1)
Ruzicka, L., Nobel Prizewinner (1939)   34(T 2.7)

Sabner, A. **42**, staff Business Manager (Publications)   42
Safety in Biological Laboratories   143
Sanger, F. **73**
  BS Hon. Member   71(T 3.16)
  Nobel Prizewinner (1958, 1980)   73
Schäfer, E. A., Professor of Physiology (Edinburgh 1899)   4
Schlossberger, J.   2
Scholarship Awards, BS   68, 69
Schools
  A level Biochemistry   170
  BS poster   169
  Careers Conferences   161, 162
  Education Group, BS   112
  Essay Competition   169
Schoolteacher Fellowships   69
Schryver, S. B., BS Chairman   27(T 2.5)
Science Museum, BS Exhibition
  BS Collection   75
  BS Exhibition   166
  BS Sir Hans Krebs Exhibition   75
Scottish Universities, introduction of Biochemistry   4
Secretaries, Honorary, BS
  1911-1945   27(T 2.6)
  1945-1986   47(T 3.5)
  nominating Committee   48
Sheppard, R. C., Peptide and Protein Group   111
Sherrington, C. S.
  founded Lectureship in Physiological Chemistry (Liverpool 1898)   6, 8
  Professor (Wayneflete Chair) of Physiology (Oxford 1914)   4
Smedley, Ida (later I. Smedley-McLean), BS Chairman   15, 27(T 2.5)
Smellie, R. M. S., Symposium Organizer   55(T 3.8)
Snow, G. A. **129**
  Chairman, Professional Sub-Committee   157
  Heads of Departments Conference   163
  Promotion Organizer   76, 129, 164, 165
Society for the Promotion of Animal Chemistry   3
Society of Chemical Industry, joint Meetings with BS (1923, 1926, 1927)   33

Society of General Microbiology, joint Discussion Meeting with BS   55
Society of Public Analysts, joint Meeting with BS (1915)   32
Spencer, B. **96**, BS Hon. Treasurer   89
Stanworth, D. R., Secretary, Biochemical Immunology Group, BS   107
Starling, E. H., Professor of Physiology (UCL 1901)   3
Students, Biochemistry
  BS Members   38, 80
  BS Travel Funds   63
Subscriptions, BS members   79, 82, 119
Swansea (University College), autonomous Department of Biochemistry (1972)   5
Symposia, BS
  Organizers   55(T 3.8)
  publication   34
Synge, R. L. M., anecdote about C.U.P.   121
Szent-Gyorgi, A., Nobel Prizewinner (1937)   34(T 2.7)

Techniques Group, BS   108
Thompson, R. H. S. **95**
  BS Hon. Member   71(T 3.16)
  BS Hon. Secretary   47(T 3.5)
  Secretary-General of IUB   152
Thorpe, W. V. **94**, Chairman, *BJ* Editorial Board (1959-1963)   123, 124
Thudichum, J. L. W. **107**   107
Thudichum Medal Lectures   107
Travel Funds
  FEBS Congress   63
  general   63
  IUB Congress   62
Treasurers, Honorary, BS
  1914-1944   26, 27, 77
  1944-1952   45, 77, 78
  1952-1986   78-89
  founder   26
Trinity College Dublin
  autonomous Department of Biochemistry (1960)   5
  BS Irish Area Section Meetings   104
  Lecturership in Biochemistry (1921)   5
Tristram, G. R., Heads of Departments Conference   162
Trustees, BS
  1929-1965   46(T 3.3)
  BS investment policies   79
  original   29
  recommendation for incorporation   45
  supplementary Trust Deed   45

Unilever Research Laboratories
  financing of Colworth Medal   66
  financing of European Fellowship   68
United Kingdom, early development of Biochemistry   3-11
University College London
  21st BS Anniversary Meeting **32**   29
  100th BS Meeting **31**   29
  Biochemical Club first A.G.M.   17
  Biochemical Club inaugural Meeting   13
  centre for development of Biochemistry   3
University of Wales *see* Aberystwyth; Bangor; Swansea; Welsh National School of Medicine
U.S.A., development of Biochemistry   2

von Euler, H. K. A. S., Nobel Prizewinner (1929)   30, 34(T 2.7)

Walker, D. G. **94**, Chairman, *BJ* Editorial Board (1969–1975)   126
Walker, R. T., Nucleotide and Nucleic Acid Group   108
Wallis, R. L. Mackenzie, Lecturer in Physiological Chemistry (Welsh National School of Medicine 1910)   5
Warren, F. L., BS Hon. Secretary   47(T 3.5)
Warwick Court, no. 7, BS Headquarters **41**   41, 81
Watts, D. C. **95, 167**
   BS Stands at international Meetings   166–168
   Chairman, *Clinical Science* Committee of Management   136
   feasibility study for *BS Transactions*   137
   Managing Editor, *BS Transactions*   138
   organizer of FEBS Trade Fair   154
Waynforth, H. B., laboratory animal welfare, BS representative   171
Wellcome Trust
   Award for Research in Biochemistry Related to Medicine   67
   travel funds   153
Welsh National School of Medicine
   Biochemistry in Physiology Course (1894)   4
   Chair of Biochemistry (1956)   5
Welsh Universities, introduction of Biochemistry   4
Whelan, W. J. **47**
   BS Hon. Secretary   47(T 3.5)
   BS International Secretary   47, 154
   formation of FEBS   154
Wheldale, Muriel, elected BS member (1913)   15

Whitley, E.   8
   Editor, *BJ*   18, 20
   endowed Whitley Chair of Biochemistry (Oxford 1920)   8
   financial support to found *BJ* (1906)   8, 18
Wieland, H. O., Nobel prizewinner (1927)   34(T 2.7)
Wilkinson, H., foundation of Colworth Medal   66
Williams, R. T. **54**
   Pharmacological Biochemistry Group, BS   102
   Symposium Organizer   54, 55(T 3.8)
Wolf, C. G. L., BS Chairman   27(T 2.5)
Wood, E. J., Education Group, BS   112
Wootton, I. D. P., *BJ* Editor   120
Work, T. S. **93**
   BS Chairman   47(T 3.4)
   BS Hon. Member   71(T 3.16)
   Deputy Chairman, *BJ* Editorial Board   121
   proposer of Hopkins Memorial Lecture   64
Worley, F. P., original BS member   **32**
Writing a Scientific Paper   143
Wye College, venue of Harden Conferences **70**   70

Young, F. G. (Sir Frank Young) **51**
   BS Chairman   47(T 3.4)
   BS Hon. Member   71(T 3.16)
   BS Hon. Secretary   27(T 2.6), 120
   formation of Biological Council   51
Young, L., BS Hon. Secretary   47(T 3.5), 156